the McCall Body BALance METHOD

the McCall Body Balance METHOD

Simple Concepts for Ageless Movement

By
Lisa Ann
McCall

Brown Books Dallas

The McCall Body Balance Method
Simple Concepts for Ageless Movement

© 2001 By Lisa Ann McCall
Book design by Jennifer Polavieja
Printed and bound in Korea

First Printing, April, 2001
ISBN 0-9706085-0-0
Library of Congress Catalog Card Number: 00-193516

BROWN BOOKS
16200 North Dallas Parkway, Suite 225
Dallas, Texas 75248
972-381-0009
www.brownbooks.com

To
My Friends Sally Ann
& Wanda Lou

Acknowledgments

It takes a world of people to create something special, and without these individuals this book would be only in my dreams. To them I say 'thank you' for making my dream come true.

The Roots: These are the individuals that built into my life for many years, a foundation to create from.

Parents, Preston and Bo McCall
Sister, Kathy McCall Laheen
Friends, Newton & Elizabeth Burns

The Creators of the Creation: It takes special talents to bring an idea to life. These individuals gave from their heart and made the magic happen.

Graphic Designer, Jennifer Polavieja
Creative Editor, Valerie McDougall
Medical Editor, Bryan Williamson PT
Photographer, Phil Hollenbeck
Illustrator, Hugo Westphal
Publisher, Brown Books

Special thanks to Angie Thusius, creator of Kentro™ Body Balance, which is the foundation that *The McCall Body Balance Method* is built upon.

Behind the Scenes: These are the individuals who gave of themselves to make this book happen. My unsung heroes.

Dr. Jim Montgomery, Dr. Lessa Condry, Laura and Doug Wheat, George Moussa, George Muller, Carolyn Lewis, Terry Caracuzzo, Fran Cashen, Ann James, George Roberts, Ireen Hosford, Ginger Bear, Jody Baird, Robert Kane, all the individuals who showed up to Friday, Saturday and Tuesday classes over the years.

My life belongs to the whole community, and as long as I live, it is my privilege to do for it whatsoever I can. I want to be thoroughly used up when I die, for the harder I work, the more I live. I rejoice in life for its own sake. Life is no 'brief candle' to me. It is a sort of splendid torch which I have got hold of for the moment, and I want to make it burn as brightly as possible before handing it on to future generations.

—George Bernard Shaw

Be smart...

This book is only for those who are going to be smart about their health.

If you have musculoskeletal related pain, get a thorough evaluation by a qualified health care professional before practicing the movements in this book.

Do not use the information in this book as a substitute for medical treatment.

Follow the instructions in this book carefully.

Consult a quality health care professional if you develop more pain or experience increased duration of pain, headaches, increased weakness or numbness in your arms or legs.

This book is designed to make you feel better, not to push through pain. Be smart and have fun!

Table of Contents

Preface

It's time we begin to understand some body truths. A natural progression of change does occur as we age but our structure is not only designed to last but to be enjoyed throughout our entire life span. So let's get back to the basics. Much of what fitness gurus and medical experts have been preaching on how to regain flexibility, strength and balance just has not shown adequate results. If the premise is wrong then the results will be as well. A big problem is we don't move enough, true, but a more important truth is that we have forgotten how to move. Realistically, as technology continues to advance, we may be moving even less. We can't afford to waste one moment of movement. Now you have a choice. Break down at 40, or run as well or better than you did when you were 20. Have a hip replacement at 75 and a blue sign hanging on your rear view mirror, or work all day in your garden.

This book shows how a person who has lived with pain for years can find relief, simply by re-learning how to move in ways that reduce the pressure on weak and painful tissue. *The McCall Body Balance Method* is not the next exercise trend. It's a re-education process that allows tissue self-healing through elimination of damaging movements and postures. It's a "body style change." So read on, and give your body a chance to work as it was intended to in balance.

CHAPTER I
MOVEMENT IS LIFE

"We usually think that what we believe is real is based on what we see and hear. But this is only partly true. It is just as true to say the opposite: that what we believe is real decides what we will see and hear in our world." [1]

CHAPTER I

MOVEMENT IS LIFE

"12 . . . 13 . . . 14 . . . 15 . . . finished." The last set of leg presses on an already tired painful knee. It had been a long, painful process both emotionally and physically. Four months earlier, Ruth had leaned too far forward in an effort to get an impossible shot, and the acquired position forced her right knee to twist with a resulting "pop." Immediate pain and swelling followed. Her knee then completely gave out when she tried to take a step. A week later, the MRI (magnetic resonance image) confirmed a severed ACL ligament, the tissue that keeps her lower leg from shifting forward at the knee. Then came reconstructive surgery, and the orthopedic surgeon said that with hard work and lots of physical therapy, she would be back in action on the tennis court in less than six months. The fear of losing the ability to play the game she loved so much drove Ruth through the fatigue and pain of her rehab.

Ruth was concerned, though. Evidently most people are through rehab and back to their particular sport in six months. Ruth still hadn't regained all of her knee motion and had developed pain in the left knee, which she hadn't hurt. The pain was probably due to compensation—Ruth was favoring her right knee and depending upon her left knee as she walked or climbed stairs.

She made a concerned call to her surgeon, and the obvious frustration in her voice prompted a change in her recovery routine. "I want you to see this lady I sometimes send my patients to," suggested her surgeon. "She has an approach that works on your body as a whole, not just your knee."

Ruth was frustrated with being shipped to yet another person who was supposed to have all the answers. "Trust me," replied her doctor, "if I didn't feel she could help you, I wouldn't send her my most difficult cases. She calls it *Body Balance* and you have to experience it to understand and appreciate its life-changing benefits."

What occurred with Ruth through the *McCall Body Balance* principles seemed almost unbelievable to Ruth. In a matter of a few short weeks most of her knee pain was gone, and all of her knee motion had returned. Simple activities such as ascending and descending steps, painful just days before, now were accomplished with ease. Most important, the thing that Ruth thought was lost forever was now within reach. This morning and just three short weeks after beginning the *Body Balance* system, Ruth was picking up her tennis racket and walking back on the court to do the one activity that she so recently felt was out of reach, maybe forever. She was returning serves, putting pressure on her surgery knee, moving fluidly side-to-side and smiling . . . on the outside, and the inside.

Perceptions of Health

Our perception of health is much like our perception of most things. If it doesn't work, buy a new one. In this time of built-in obsolescence, we think we can replace anything, and this is affecting the way we approach our bodies. Sure, there are times when we need to replace a hip or a knee, and we're fortunate to be able to do this. Our misplaced worry is about the replacement cost, thinking "insurance will pay for it." By so doing, we completely overlook the real cost to our bodies and our psyches chalked up by not being in tune with our bodies. We start worrying that lifting a trashcan or carrying luggage will cause pain, so we become "victims" of our bodies. Yet we go to a gym and see how much weight we can lift.

We become dependent on our environment to meet many of our needs, and when it does not, we are helpless. For instance, we ride in luxury cars, get food at drive-throughs and talk to answering machines while being entertained by TV. We depend on the outer world to meet our basic needs. Yet we know that a long walk or watching a sunset makes us feel better and adds a little magic inside. We need that magic, but have forgotten what we used to do as children and young adults to feel alive. Think about it. When you were a child, were you happy when you were forced to sit still, or did your happiness come with motion?

New Views of Movement

Movement is life. It's that simple. Every cell literally maintains life by moving and re-creating itself in the process. Every cell depends on movement to exist: blood pumps through the chambers of our heart; our lungs transport oxygen to the blood providing nutrition to keep cells alive. Our thoughts send messages from the control centers of the brain to change our mental perception and physical responses toward life.

Movement Is a Language

Body language is more than what people reveal about themselves through their physical display. It is a communication that enables the body to stay in tune with itself and in harmony with gravity. However, we are losing the ability to know ourselves on a very basic level. Almost unnoticeably, lifestyles have become sedentary. Our body becomes aware that normal activities of the day are no longer easy. People find getting in and out of their car, or sitting comfortably in a chair, is more difficult. This is old age. It has nothing to do with chronological age but the dying off of a physical principle of life: natural movement.

Moving naturally does not need to be learned. Look at babies as they develop, sitting, bending or crawling then standing, walking and running. All they need is an environment that allows them to move as they were created to move. These small physical steps build on the "body intelligence" gained the day before, and the day before that. The fact that most people suffer back problems sometime in their life helps illustrate that most people lose their "body intelligence" in childhood. Too many of us unlearn this natural body intelligence by following the incorrect models we see all around us.

We do not have a template of how to move in Western culture that makes it easy to move without worry, no matter what our age or activity. This means we do not have guidelines on what the body needs to know to maintain healthy, fluid motion. Current technology allows scientists and human

Yet we are still at a loss on how to sit in a chair and get up and down from the couch because over time, with "practice," we get worse at it, not better.

performance specialists to analyze the precise positioning of a cyclist's pedal or each small fraction of a golfer's swing. Yet we are still at a loss on how to sit in a chair and how to get up and down from the couch because over time, with "practice," we get worse at it, not better. We skip the most vital step that influences all movement: how to move moment by moment in our environment. If we do not know how to adapt and change, we will continue in the state we are today—constantly at the mercy of the unknown and vulnerable to unnecessary injuries. This limits our enjoyment of the most basic things in life. This is not the case in many cultures where people's way of life complements their physical experiences of living. We spend more time fixing our lives than living them.

I'm not advocating a new set of rules because rules may constrict our creativity. I'm not suggesting a list of "do nots." This book says we do know the language of body movement, we have just forgotten how to speak it. *McCall Body Balance* restores this language and gives us back what we had as children. It is time to create an environment in our daily activities and lives that brings this intelligence back and allows us to move and live at home in our bodies ... again.

McCall Body Balance is a way to reeducate ourselves on how to move and how to think about movement. The body

Consider this

1. *You never have to hold your stomach in again* and your back will be better if you know how to move correctly.

2. *You strengthen your abdominals by using your whole torso or trunk.* Simple movements such as walking and climbing steps, getting in and out of chairs, bending and turning, all strengthen the abdominals if you know how to move your spine on your pelvis correctly.

3. *A comfortable car ride can help correct many common back problems* when you know how to sit.

4. *When you carry a purse or luggage on your shoulder, the upper back becomes more upright and stronger* when you learn how to apply *Body Balance*.

is already wired to move well. To get the flow back, you must understand how to make the simple acts of standing, sitting and bending (plus all those "mindless" movements in between) work for you so your joints will last longer than your appliances.

What Is *McCall's Body Balance?*

Optimal posture, as proposed by the *McCall Body Balance Method*, is an exception to the rule. It challenges that the dominant perspective of posture held in the U.S. for more than 50 years is wrong. *McCall Body Balance* is based on my observations, interviews and practical application of this method. It's a first attempt to get back to the basics by taking a broader view of posture, studying those peoples across the globe who move the best, then building on this fundamentally different understanding of movement.

McCall Body Balance is the process of relearning the most basic components of all movements, which are built on what I call Universal Movements. Universal Movements are the simplest elements that make up all motion. These moves include standing, bending, sitting, turning, lying or walking as well as the transitions in between. Universal Movements are the foundation to the more complex moves you would make on a golf course or in a bowling alley. In more developed countries, people have lost the understanding of the Universal Movements. *Body Balance* brings back this lost knowledge.

The famous dancer/choreographer, Martha Graham, speaking of the art of dance, put it this way: "All that is important is this one moment in movement. Make the moment vital and worth living. Do not let it slip away unnoticed and unused."[2]

Judy's Story

Judy, a 60-year-old, came to *Body Balance* complaining of multiple joint pain and limitations. She was unable to get in and out of cars well or move from her chair without problems and pain. She had poor balance when walking and getting up from the floor was impossible. She had osteo-arthritis, rheumatoid arthritis and ankylos-ing spondylitis (rarely found in females). To put it simply, her joints were shot. Her history basically amounted to back, hip, neck and "you name it" pain. My heart went out to her for all she had endured and I asked her why she had waited so long before doing something about it. She leaned over my desk and whispered: "I thought it was normal to hurt as you got older; I'm just at the point now that I can't move!"

When she came to my *Body Balance* class, I'd help her get up and down from the floor. The day came when I watched her use these principles, and slowly, with great concentration, she stood up from the floor on her own. And more important, she stood up without pain. She raised her arms in the air as if she'd just won Olympic gold and shouted: "I DID IT!" A small miracle had just occurred and she headed home to show her husband, children and grandchildren. As time passed, she integrated more of the principles and was able to take a trip to Europe. To her amazement she did very well with the walking and traveling: a quantum leap for a woman who at one time could barely move. It all starts with the smallest victories. The little stuff matters.

McCall Body Balance teaches us how to regain correct Universal Movements. We learn how to break moves into their compo-nent actions. We then practice focusing on these components before finally integrating them back into all activities, from the sim-plest to the most complex.

Body Balance is a skill, a science and an art. Skill implies "superior performance along with greater repeatability of performance."[3] Science is defined as "a systematized body of knowledge based on observation, and experi-mentation."[4] Art is a form of self-expression or creativity. In this case, the art of body bal-ance is expressed through movement.

Most people in our culture do not move with great ability or proficiency (skill). As they move across a room or get up from the floor, they do not represent a systematic form of applied knowledge (science) or a very creative expression of their self (art). It is a mess!

Many people comment on how much they enjoy being able to get up from the floor with ease after learning the skill, art and science of movement. It may seem ele-mentary, but try getting up and down from the floor about 10 times in a row. Unless you are a young, conditioned athlete, it will wear you down and your last attempt will look and feel much different from the first! Many people do not get down on the floor because it seems difficult. They avoid it because it uses energy, which is limited for most of us.

A person may notice slight limitations in simple acts such as reaching down to tie a shoe or bending to get the paper, without realizing these limitations have a snow-balling effect. One day, when they pick up the paper or swing a golf club, all move-ments stop due to a major breakdown. *Body Balance* allows the simplest movements to create the greatest changes in our body because each move builds upon another.

Posture is not just to please mother

If pain relief doesn't motivate you, maybe discovering a more attractive and comfortable body will get your attention.

People are extremely conscious of the designer, the style and the price of a silk suit while what's supporting the silk suit is often forgotten. We may care about the fat, but we don't talk about the structure.

I'm sure you've been at a party and noticed the girl in the little black dress. All that work at the gym can be undone if you can't move well in your "birthday suit"—let alone in a business suit.

It comes down to posture. Even hearing the word sends shudders up my spine! Chin in, chest out, stomach in . . . head up . . . straighten those shoulders . . . stand up and pull up. I get tired just remembering the phrases of my youth, let alone trying to carry them out. When you try to stand the "right" way, it wears you out, but relaxing makes you sag without a struggle. It's that meaning of "posture" I want to bury.

However, the idea of standing properly without a struggle is actually correct. Why are we the only animal on the earth who has to work at standing correctly—yet still struggles to do so? Don't give me that jazz about "if we were still on all fours, movement would be easy." I am not a "victim" of standing on two feet and neither are you.

The human body was well designed for the spine to balance on the pelvis, steadily and strongly. Look around you and see if people struggle with moving. Do they sit with style and grace or are they more inclined to flop into a chair and crawl out? They often look like they will barely make it to wherever they're going and when they get there, what a relief!

> Sitting as the body is designed brings out the best in any body.

I'm sure you've been at a party and noticed the girl in the little black dress. All that work at the gym can be undone if you can't move well . . .

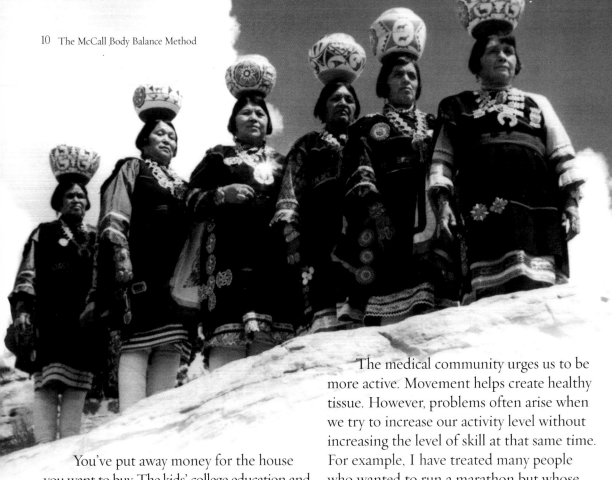

You've put away money for the house you want to buy. The kids' college education and the retirement plan are a reality. Now apply the same concept to movement. There are certain movement principles that must be known so you can swing that golf club in your golden years. The reality is that joints will only work when they move as they were designed to move. How golden will those years be if your body has lost this understanding of its design? Why are we ignorant about something we learned instinctively as children yet do so poorly as adults? It gets back to our perspective again. Having money, time, freedom and health creates the illusion that life will go skipping along in a carefree way. If one of these luxuries disappears, then things are seen and valued in a drastically different way.

> The old joke that gravity is working against us as we age does not have to be true if it is allowed to work as a positive force.

The medical community urges us to be more active. Movement helps create healthy tissue. However, problems often arise when we try to increase our activity level without increasing the level of skill at that same time. For example, I have treated many people who wanted to run a marathon but whose bodies broke down because they did not have the skill to increase their running distance. On the other hand, I have worked with inexperienced runners who faithfully did the *Body Balance* moves, then finished their marathons because they improved their skill level as they increased their distance. Yes, we need to be more active, but we have to start by getting the fundamentals right—daily movements—and build from there.

"Daily movement" needs to be redefined. These are Universal Movements, the foundation of all movements. If they're done incorrectly, over time the cost will be the health of the joints and the tissue that supports them. These have been termed repeti-

> . . . You should be able to carry 20 per-cent more weight, the groceries or luggage, without spending more energy or effort.

tive stress injuries by the medical community.[5] Over time, with the many repetitive movements of daily activities performed incorrectly, the tissue wears improperly and sets the body up for multiple joint problems. When the fundamentals of correct movement are practiced, a new awareness arises as "movement" takes on a new and better meaning. The old joke that gravity is working against us as we age does not have to be true if it is allowed to work as positive force.

There are people in the world who really make gravity work for them. Critical research in this area was done by C. R. Taylor and N. C. Heglund.[6] They studied certain African women as they walked and carried items on their heads or backs, their traditional carrying methods. Amazingly, they could carry an extra 20 per-cent of their body weight without expending a single additional calorie! In other words, the increased load did not cause them to use more energy; they basically carried the

Good Posture it's Your Body's Savings Plan

weight "free." How do they do it? Their secret is an "energy-conservation mechanism," according to the researchers. Part of their trick is using their bodies more efficiently. Applied to your life, that means you should be able to carry 20 percent more weight, the groceries or luggage, without spending more energy or effort.

Think of your body as an inverted pendulum system. When walking, your energy is transferred from kinetic energy (energy in motion) when both feet are on the ground, to gravitational potential energy (stored energy for use) when standing on one leg. In a perfect pendulum there is no wasted energy. In most humans and animals the exchange is only about 65 percent. African women increase the efficiency of transfer as they carry heavier loads, compared with European men and women in the study who did not.

It is not realistic to suggest carrying shopping bags on our heads. In the grocery stores, instead of "paper or plastic?" the line

could be "head or back?" We can learn to move as the Africans move and greatly increase our efficiency at carrying things on our shoulders or backs as well as at the sides of our bodies.

Building Physical Intelligence

We all know you cannot be good at something without practicing it. This is also true when carrying bags, swinging a golf club or carrying a child. As with sports, the skill comes with the practice. Once we learn a sport that demands skill, or challenges us to maintain a specific posture, we become more efficient in that movement and the movement becomes more interesting.

Take snow skiing. Your instructor teaches you the basic principles of how to go down the hill without losing control, and how to stop. The first time down the hill you may feel awkward, but you follow the directions or you know you could hurt yourself. Every time you go down the slopes you keep the basic principles of skiing that you learned the first day. After a while the movements become more natural and your physical awareness of the skis and their relationship to the slopes

becomes second nature. You still stick to the basic principles you learned the first day. If you decide to change the fundamental principles of movement and begin to tuck your buttocks, stand straight and lock your knees, it would result in a disaster. A large mountain and a pair of skis gives a person tremendous respect for correct movement.

Now apply this principle to what many consider mundane movements. While getting in and out of a car or a chair is less exciting than tackling a mountain, the importance of these subtle moves may not be noticed until you are speeding down life's "time" mountain. We must build on sound principles of movement.

Many patients complain that traveling physically tears them up, saying "All the lifting and sitting on planes kills me." In reality, traveling is no more injury-stressful than movements we do daily to get to and from work and around town. We use the same movements of everyday life. By relearning the basics, travel can be a pleasant experience.

A large mountain and skis give a person tremendous respect for correct movement.

Back
Back School
School

This man understands how to use his body. His low back is built to carry weight. He is getting something accomplished while doing his "weight lifting workout." What a novel idea!

In the medical community, we have popularized the concept of "back school."

Many of my patients tell me they have gone to back school and say, "I know how to move correctly to take care of my back." That phrase "know how to take care of my back" bothers me because we are separating our backs from the rest of our bodies. We're talking about our back as if it were an extension of ourselves rather than an integral part of ourselves. We've fallen into the "specialist trap" where we see one doctor for one part of us, another for an adjoining part and treat our body as if it's not an integrated whole!

The low back is the center of where movement begins. It is the hub.[7] This area should be viewed as the strongest portion of the body—not as the problem but the answer to the problem of incorrect movement. If "back school" works for an individual, then every day after the first lesson, that individual would become functionally more aware and movement would improve, just as the snow skier improved on the slopes year after year. So, five years of back school would be similar to skiing down the most difficult slopes because correct movement builds upon itself. Anything that truly works builds upon itself. It is obvious when watching a child progress with movement. The complexity of the skill begins with stabilizing the core of the body when sitting upright, then bending to stand, then developing more skillful moves such as skipping, running and riding a bike. The body was designed to learn movement through a smooth progression of simple to more complex.

> The low back should be viewed as the strongest portion of the body; not as the problem but the answer to the problem of incorrect movement.

The Amount of Change Depends On You

When you bend at your hips correctly, your body will become more in balance, more flexible, more relaxed and stronger than it was when you started. Body Balance moves expand your energy reserves so you can enjoy movement.

Body Balance is not about gaining perfect alignment but about the process of moving your body in the right direction so that in time it will move more and more the way you want it to move. It does not matter if changes are small or large. The amount of change is up to you and what you want out of movement. Remember, the body has an amazing capacity to adapt. If you learn to play the piano, in time your fingers will remember how to play the scales and eventually more difficult compositions. This is called a "motor engram." Within time, your fingers and mind work together to play a melody. The body and daily movements work the same way. "If you don't use it, you lose it" is a phrase that fits movement well. When joints do not move through their natural range of motion, compensatory changes occur that lead to breakdown in tissue. This problem leads to other joints malfunctioning by moving in ways they were not designed to.

If you bend incorrectly, like Phil, you will be vulnerable to more and more injuries.

Doris's Story

DoriS's Story

Doris is 76 years old. She came to *Body Balance* suffering hip pain, which made it difficult to get down and up from the floor. She's an avid gardener and wants to stay that way. She thought her gardening days were over because of the pain in her hips. Now she works in her garden for hours and hauls fertilizer, mulch and grass by herself because she learned the key to bending, and she gets better at it the more she does it. Her exercise program was to move correctly as much as possible, and now it is second nature. Her younger neighbors saw her working long hours in her garden and asked for her "secret."

The Physiology of It: The Body Must Move As It's Designed or There Is Breakdown

There is a communication breakdown in the neurological system because of this movement mix-up. If a joint is not moving as it should on a day-to-day basis, it cannot gather the information

through the senses to send the desired message to the brain for the desired responses.

For example, if you close your eyes (a sensory receptor) you cannot see the external stimulus (a flower) to interpret what it is and create the desired response to pick, or just admire, the flower in front of your eyes. The joints of the body and the skin have thousands of receptors with the potential to tell you much more than you can imagine about your external environment and how to respond to it. The connective tissues surrounding these joints and skin "have their eyes closed" due to lack of correct stimulus. This can change with *Body Balance* because correct stimuli through movement can allow you to "see" the world again.

If we continue moving improperly, we move blindly and without freedom in our movements throughout the day. Those movements often create pain, which means tissue damage is already occur-

ring.[8] This loss of movement becomes more serious since pain creates more restrictions through muscle guarding or spasms. At this point, joint dysfunction has occurred on multiple levels. If the hips do not move well, joints such as those in the knee and back are vulnerable to unnatural stresses or forces. This can cause problems to the neck and shoulders or the feet and ankles.

This domino effect does not have to happen. A movement as simple as bending forward incorrectly can create a multitude of problems. On the other hand, when bending is relearned the natural way, it makes many changes in the right direction. The game of golf, for instance, would not

> If we continue moving improperly, we move blindly and without freedom in our movements throughout the day. Those movements often create pain, which means tissue damage is already occurring.

create a club of back sufferers if golfers moved properly. The spine loves rotation.[9]

Gravity plays an important role in our lives. It plays a significant role in helping the structural part of our body stay healthy. When gravitational force falls upon the correct landmarks of the skeletal structure, it enhances the body's ability to move and increases bone strength. The growth and strength of bone is described in Wolff's law: "Bone will change its internal architecture according to the forces placed upon it."[10] "According to the forces placed upon it" is a vital statement. If the forces placed on bone by gravity are incorrect, there is a breakdown in different types of tissue, bone being one of them. We see bone as static, but it is alive, changing and adapting with all our movements. For instance, without the formation of new cells a broken thigh (femur) bone could take 200 to 1,000 years to heal.[11] There must be stress on bone to stimulate the cells to regenerate and keep the bone renewing itself.[12]

Like all biological tissue, bone can break down under forces that create degenerative changes, especially the major weight-bearing bones of the body—the spine and hip. Lack of correct weight on bone contributes to osteoporosis.

Movement needs to be understood in an integrated way. Because we are not comfortable carrying heavy loads, we roll our luggage through the airports. We think carrying a heavy purse on our shoulder will create structural problems and strain muscles, yet when it's done in an integrated way, it actually does the opposite,

If these people do not carry their "luggage" as they do they will fall into the modern trap of machines replacing human movement. This in turn would rob them of a healthy spine and back.

I call these "movements in the moment." They are seen as boring because we do not have the physical know-how to enjoy them. Once you begin to integrate *Body Balance* into your life, moment-by-moment movements become enlightening. It has been well documented that one of the reasons people are mentally drained is that they rarely live "in the moment." Too often we think about the past or future. But why not make "movements in the moment" better—because that is where life is spent.

All of us want to be outstanding in our favorite activity or sport. We also want to stay physically attractive into later life. The most common visual "aging" problem I see is not gray hair and wrinkles but the loss of an upright, healthy posture. Many people think stooping is inevitable along with the loss of balance and stability, but other cultures have shown this is not the case. These changes are often subtle and not recognized in the beginning because they build on themselves just like improvements in *Body Balance* do. People look younger when they move well.

Understanding the fundamentals of movement takes all other movements to a higher level.

Movement needs to be understood in an integrated way. Because we are not comfortable carrying heavy loads, we roll our luggage through the airports.

building upper body strength and counteracting osteoporosis. The nonintegrated way is to not carry that heavy purse, then go home and do "light weights."

I live in Dallas where one of the most popular sports sidelines many people. That "sport" is shopping. One of the greatest compliments I receive from patients who love shopping is hearing their pleasure at being able to walk through malls for long periods of time without feeling pain or discomfort in their joints.

People do not want to move or exercise if they experience pain during the most natural of all activities, walking. Daily movements are a forever thing in life. You won't be hang gliding or boat racing or bungee jumping forever, but you will be bending, sitting, standing and walking, and you'll want to do them well as long as possible.

Movements in the Moment

Environmental Hazards

These days, we realize the huge impact our environment makes on our health. We began addressing pollution in the 1960s. In the past few decades, we have studied heart disease and diet and their relationship to our environment, a lifestyle of fast food and overeating. The same holds true for the way we move or do not move. We are the most mobile immobile society. We can travel around the world with little physical effort, yet carrying out the trash causes us mental and physical anxiety. The laws of nature make it clear that we must move. It is vital for our physical health. It is what keeps our joints stable and our backs strong. Without correct movement, the joints fall apart. It is no surprise that we need back braces, knee braces, wrist supports and even mouthpieces so that all these joints will hold up. They are like those $2,000 chairs that promise relief of back pain. A chair cannot fix you, but you can: with your movements.

> The laws of nature make it clear that we must move. It is vital for our physical health. It is what keeps our joints stable and our backs strong.

We look for joint comfort without realizing we are literally sitting on it. Using our body produces the changes. Learning how to walk up the stairs without pain can knock off pounds and keep knee replacements out of sight. Taking the trash out is the answer, not the problem. Immobility robs us of many wonderful experiences that could be realized if we begin to recognize the opportunity.

How We See Ourselves Through Movement

We all move and hold our bodies according to a mental picture we have of ourselves. A story about Picasso illustrates this well. When he was asked by a man, "Why don't you paint people as they really are?" Picasso replied, "And how are they really?" The man took out a photo of his wife and said, "Like this." Picasso responded, "She's awfully flat and small, isn't she?"[13]

What is attractive is very much influenced by our perspective. Our cultural ideas of what is physically attractive have been greatly influenced by the media: advertising, TV, billboards, magazines, etc. This influences the way we hold ourselves and how we move. (We shouldn't need to "hold" ourselves. We should be relaxed and natural.)

Terry's Story

Terry came to me with long-standing knee pain. He loved running but, despite surgery on his left knee, he was still in pain when he ran. Many years passed and this chronic knee problem kept him from running. There were obvious alignment problems. His legs were bowed (severe genu varus), which put an undue amount of unnatural stress on his knee, plus it constantly irritated his hamstrings and other muscles. His pelvis was in a posterior tilt (tucked buttocks), which created compensatory alignment problems in his rib area (mid-thoracic) as well as even more stress on his knee due to internal rotation of the thigh bones (femurs), just to mention a few of the specifics.

The first challenge was to reeducate his body. *Body Balance* therapy changed the stresses through his legs and influenced the angle of his pelvis. This, in turn, allowed his trunk to be supported more easily on his pelvis and decreased the stresses around his knee. A "reshaped" Terry now had a far more fluid running movement. After a dedicated few months, changes occurred but they occurred on more levels than his knee.

When he was back in stride as a runner, Terry thanked me for fixing more than his running ability. He told me that his self-image had taken a huge shift for the better. He was 48 and for the first time in his life, he felt comfortable in his body. This seemed to put his knee problem in a different perspective. He can now run up to six miles before any discomfort arises. With his knee much better, Terry is functioning mentally and physically on a much higher plane. Some of these changes, for Terry and others, can be profound.

To be healthy, think about feeling good then looking good rather than being a victim of society's standards by putting looks first. If feeling good influences our movements first, this will naturally flow into looking good. If we act in that order, new ideas of how to move will not be so difficult to accept.

The idea that movement can change the way people view themselves may sound a little far reaching, but it's logical if you think about self-image.

I have seen joy in the faces of my patients when they began to understand how to move. It was more than getting out of pain. They went beyond what they expected and even thought possible.

It does not matter how old we are, our body always responds to correct movement. It does not live by numbers but by experience.

We accept the fact that back pain is normal, but what this really shows is that we do not understand how to move intelligently. Even the medical books today say that having back pain some time in life is normal.[14] To say something is normal confirms its reality. However, a new, healthier perspective is on the horizon—a new version of normal.

Movement, and the ability to do it well, should be part of every aspect of our lives. To regain this, we have to start by understanding the body as it was designed, then expand on that to create an everyday intelligence for everyday movements. The key is getting physically involved with understanding the how. Knowing how to move transforms everyday moves into something of art: grace.

This beauty shop is free of fancy chairs. These women sit in balance easily with no more than a rock for support.

Lisa Ann

My Haptic Ways

Research shows people learn in seven ways. Some learn by sight, others through sound. My learning mode is through movement. It wasn't surprising, therefore, that my chosen field was all about movement— physical therapy. It was a matter of moving forward to take what I'd learned to the next plane.

Physical experiences began molding me very early in life. At nine, I was dancing for my parents' bridge club. For years, my sister played the piano while I danced for imaginary audiences. It seemed totally normal to me to be dancing on the picnic table at a birthday party in 5th grade. If you tied my hands, it would probably impair my speech. I hear what people say when they move. This means I'm a haptic type; I discovered the world through my sense of "feel" or kinesthetic experience.

Viktor Lowenfeld, an art teacher, states: "The haptic type utilizes muscular sensations, kinesthetic experiences, touch, impressions, and all the experiences of the self to establish his relationship to the outside world."[15]

It is the kinesthetic expression that gives meaning to motion. The artful dancer's spin or the poetic skill of the athlete in motion draws the eye and maintains the observer's interest. The size, strength or flexibility of an individual is nothing without coloring the movement through the kinesthetic influence.

Over years, many of my achievements have been in the physical domain. I ran a couple of marathons, and spent five years competing in triathlons. All the time I gathered new ideas, some injuries, and refined the things I saw that enhanced movement.

oBServe

When you go out today, observe how people move. Do you think we move with skill and grace in our society?

1 J. W. Hayward, Letters to Vanessa, Shambhala, Boston, 1997, p. 7.

2 Jon Kabat-Zinn, *Wherever You Go There You Are*, Hyperion, New York, 1994, p. 45.

3 B. Brownstein and S. Bronner (eds.), *Functional Movement in Orthopaedic & Sports Physical Therapy: Evaluation, Treatment & Outcomes*, ed., Churchill Livingstone Inc., New York, 1997, p. 231.

4 *Webster's New World Dictionary & Thesaurus*, 1996.

5 Ola Grimsby, Clinical & Scientific Rationale for Modern Manual Therapy, MT-1, San Diego, 1998, *Exercise Physiology*, p. 6.

6 C. R. Taylor & N. C. Heglund, Freeloading Women, *Nature*, 4 May 1995, p. 375.

7 Brownstein & Bronner, 1997, p. 145.

8 Wyke, 1979b; Freeman and Wyke, 1967, *Manual Medicine*, Thieme Medical Publishers, New York, 1990, p. 36.

9 F. Jacobson, "Medical Exercise Therapy," *Norwegian Journal of Physiotherapy*, 1992: 59(7):19-20.

10 J. Wolff, *Gestezkter Transformation der Knocken*, Berlin, A. Hirschwald, 1884. Ola Grimsby, Clinical & Scientific Rationale for Modern Manual Therapy, MT-1, San Diego, 1995, *Functional Histology*, p. 23, cited by second source given.

11 T. Malone, T. McPoil, A. Nitz, *Orthopedic and Sports Physical Therapy*, Mosby, Missouri, 1997, p. 31.

12 J. A. Gould and G. J. Davies (eds.), *Orthopaedic and Sports*, C.V. Mosby Company, Missouri, 1985, p. 35.

13 J. W. Hayward, Letters to Vanessa, Shambhala, Boston, 1997, p. 8.

14 T. Malone, T. McPoil and A. Nitz, *Orthopaedic and Sports Physical Therapy*, Mosby, Missouri, 1997, p. 539.

15 T. Armstrong, *Seven Kinds of Smart*, Penguin, 1993, p. 82.

CHAPTER II
MY JOURNEY

CHAPTER II

MY JOURNEY TO McCALL BODY BALANCE

In the late 1980s I worked in a physical therapy clinic packed with the latest specialized equipment. Patients came to "our" environment to be treated, then were sent home with a regimen of exercises that frequently had little application to real life.

The limitations of this system really struck home the day Joan, a surgeon's wife, made the last visit of a grueling rehab session after a knee reconstruction.[1] She thought I was a saint for getting her "through" it. I felt rather proud receiving her fine thank-you gift. Then she dropped the bomb. Her family was moving back north, she said, "So how can I care for my knee without all of this equipment?"

Her comment was so innocent and honest. All she wanted to do was ride bikes with her kids and go on walks with her family. She had no interest in finding a gym with a lot of "stuff."

Her comment changed my direction in therapy.

This chapter explains my journey in developing *The McCall Body Balance Method.* It shows some of its roots and real world examples, past and present, that convinced me I was on the right track.

I'm not the first person to have noticed the frequent gap between clinic-based therapy and real life. Shaw Bronner, physical therapist, in his textbook on functional movement stated that these two worlds—the clinic and real life—seemed disconnected. Despite extensive evaluation and thorough treatment of their physical problems, he said many patients were still afraid to move for fear of pain or reinjury.

"Perhaps we continue to neglect the most important link: return to functional movement with its many directions and richness of expression."[2] *Body Balance* is the key to restoring that functional movement. It looks at movement from a different perspective, redefining it so that the two worlds come together. But, let's get back to the start.

Patients came to "our" environment to be treated, then were sent home with a regimen of exercises that frequently had little application to real life.

My experience with Joan made me face some uncomfortable questions:

* What are we really doing for our patients?
* What should we be doing for them?
* If insurance doesn't cover therapy costs, will patients still come?
* If you take all the bells and whistles (equipment) away, what's left?
* Have we segregated the body from the person? Only later did I learn that by treating the body as it was designed—in an integrated fashion—it was feasible to combine the science, art and skill of movement.

Hatha Yoga: My Path to Integrated Physical Therapy

Joan's comment and my inability to solve my own injuries highlighted the need for new ideas and a simpler approach. My search introduced me to yoga. I studied, then taught Hatha yoga while continuing as a physical therapist.

Hatha yoga introduced me to integrated movement. For instance, it holds that to treat a knee injury correctly, a practitioner must address what is happening at the pelvis and spine, not just the knee. It is important to treat the body as an integrated whole instead of a series of problems to be dealt with in isolation. Four other valuable lessons rapidly dawned:

1. Without the fancy equipment of the physical therapy clinic, brains and hands can create answers and solve functional problems.

2. Group sessions are valuable for patients because they focus on wellness, thus creating a positive attitude. Patients attend a "class" rather than a "clinic." A class lets people have fun correcting problems. It's also highly cost-effective for patients and greatly needed in these days of managed care.

3. Relaxation at the end of class is highly valuable. A key factor in the *Body Balance Method* is understanding relaxation through movement.

4. Focusing the mind on a movement creates important mental and physical changes in the body.

Yoga: East and West

" The wisdom of the mystics, of the Sufi, of the great yogis, or of the Zen masters might have been excellent in their own time—and might still be the best, if we lived in those times and in those cultures. But when transplanted to contemporary California those systems lose quite a bit of their original power . . . Ritual form wins over substance, and the seeker is back where he started."[3]

Wisdom can lose its power when transported into a different culture if we don't sort the essential elements from their setting. Yoga postures are a prime example.

The Original Yoga

Yoga, as most Westerners practice it, is a distortion of the original. Pantanjali, who lived some time between 500 B.C. and 200 B.C., is considered the originator of yoga.[4] His teaching focused on all aspects of life, from man's code of conduct to man's vision of his true Self.[5] He wrote on medicine and grammar as well as yoga. The word yoga comes from the Sanskrit root yuj to "yoke" and is the

yoking or union of one's soul, body and mind with God. By viewing God in this manner one is able to view all aspects of life evenly. [6]

There are eight limbs to this Indian science: universal ethical principles (Yama), rules of personal behavior (Niyama), the practice of Yoga poses (Asana), the practice of Yoga breathing techniques (Pranayama), controlling the senses (Pratyahara), focus or concentration of the mind (Dharana), meditation (Dhyana) and absorption in the Infinite (Samabhi). [7]

The original idea was that each limb would be built upon an earlier one and they would work together as a whole. They were not designed to be segregated. By the time devotees started on the postures, it was assumed they had addressed ethical life and personal conduct. As they excelled in the Asanas or postures, they would move into the next stage, breathing, and so on.

The Original Meaning of a Yoga Pose

In our society, the most practiced of the eight limbs is the Asanas (poses) which is to bring about balance, while making the mind still. [8] When the poses become effortless, one has mastered the Asana. This effort, concentration and balance forced one to live in the moment. An Asana acts as a bridge to unite the body, mind and soul. [9]

This is a little different from the yoga marketed today. It is a beautiful explanation of how the body should and could be a part of the way we live. I saw my challenge as discovering how people in our society could live in the moment and unite their body and mind through every-day movement.

This women must "overstretch" important ligaments in her back to gain the ability to bend in this way. She is not bending just at the hip but along the spine to gain a "pose" that has no use or purpose.

> The yoga postures, like a set of exercises, did not overlap or connect to the real world of daily movements.

Pantanjali lived in an environment much different from ours. Even more important, he and his followers were in harmony with the poses because he was building on bodies that were already in balance. Not only is our environment today totally different, but our bodies are not in balance. Thus, yoga postures are like trying to fit a square peg into a round hole because the poses require advanced movements of bodies that have not learned the basics!

However, it also became apparent that the benefit of many of the yoga moves was limited. They did not solve—for me, at least—the problem of establishing a nexus between movement and people's daily existence. While yoga offered a new perspective on the body, the yoga postures themselves did not show me a practical way of moving in a comfortable, injury-free way on a day-to-day basis. I can do a mean "downward dog" but how is that going to teach me the best way to lift my dogs into the back of my truck? People should be able to lean over and unload a case of drinks from the trunk of a car on a daily basis without fear of injury or pain for as many years as they want. A patient's ability to function should be limited by their desires, not the practitioner's skills.

The yoga postures, like a set of exercises, did not overlap or connect to the real world of daily movements. The Western way of life is different from the Eastern traditions of yoga where the postures originated.

In this yoga pose the focus is not on how each joint needs to move. By repeating this pose over time this woman can damage joints. Her elbow ligaments are actually supporting her weight, which they are not designed to do, and her knees are locked out, putting improper pressure on the joint.

Incidentally, it was fascinating to discover that with *Body Balance*, I no longer needed yoga postures; I had gained the connections to bring the physical and mental together. Besides correcting my injuries, I am able to relax (a major accomplishment) and be in the "pose" of the moment.

Back to Basics

About the time I was trying to integrate yoga and daily movement, I read an article in *The Yoga Journal* on the study of movement on an international basis.[10] Angie Thusius described how people in other parts of the world move well without joint pain and dysfunction, regardless of age. This different approach to movement immediately struck me as another part of the answer. Angie had been introduced to this idea of cultural movement by Noel Perez, a woman who spent many years researching the movement of people from different ethnic backgrounds across the globe. Angie expanded on Noel's studies and in turn developed her body balance guidelines to help restore natural body movement.

This insight on movement highlights the fact that the "correct" way to move in Western cultures is not regarded universally as the right way. A continent away, ordinary people are moving well, without pain, using different principles of movement. The principles I learned from Angie were not any one person's philosophy or belief but a universal norm all people can experience and claim as their own. In reality, it is the natural way all people begin to move as children. I saw this as true in our own culture. Children until school-age move with the same principles that other cultures do throughout their life.

When applied to my knee restoration patient, it became obvious that the spine had to sit on the pelvis correctly and move in harmony to correct the knee.

If people understand how their bodies work through their daily experiences, they will have the mental and physical confidence to reclaim their natural body wisdom. This gives each of us the ability to have good somatic judgment and know when there is something wrong in our body before it gets out of hand. This wisdom is the ability to correct a problem at that moment, helping us to stay in the moment.

. . . the "correct" way to move in Western cultures was not regarded universally as the right way.

Children until school-age move with the same principles that children in other cultures do throughout their lives.

Reality Check

On the plane back to Texas from my first lessons with Angie, I knew I had found some answers, yet could only think of more questions. Some were big. How do we reach people across the country who do not understand the impact their daily movements have on their bodies? Some were more personal. How would this work fit into my physical therapy practice? Or, how in the world was I going to get people in Dallas, Texas, let alone the rest of the United States to see that relaxing their bellies (which tend to be surgically tucked) would make them feel better, look better and live healthier? Fortunately, I've learned to follow my intuition rather than my fear of being unemployed or of looking stupid.

The transition into *Body Balance* began. I lost a few followers and gained a few converts. Most of all, I became more fascinated by the dramatic changes in my patients' bodies, and my own. My alignment and posture now makes moving easy. My neck pain from a cycling accident, my hip problems from a stress fracture while running and all the other injuries I had collected over the years were being corrected as I learned to relax into the natural movement patterns my body had "forgotten." When I took these principles into the water, I found that during swim workouts, my back and abdominals were getting much stronger because I was moving more efficiently. This toned up my body as I gained faster swim times.

Passion and insight have more power than doubt and rejection. I had the passion and was gaining more and more insight on how the body really worked. I could not explain it on a scientific level then, but many of the typical joint problems that the orthopaedic community continued to scratch its collective head about could now be rehabilitated more effectively. I now knew by observing people in other parts of the world we could find answers that our highly skilled medical community lacked.

HOW Do otheRS in the World MoVe?

I now knew people in other parts of the world had answers that our highly skilled medical community lacked.

Our Earth is full of people whose ideas of movement differ from ours. This encouraged me to learn more in my quest for the universal norm of movement.

Janet's Story

Janet's Story

Janet grew up in a small, central African village until she went to school in England at age 18. She's lived in the United States for five years and understands the different lifestyles and their influences on her body.

In the village in Malawi where she grew up, all transport was on foot and "modern facilities" were absent. She showed me that her people bend deeply from the hips with a very straight spine. They carry heavy items on their heads and backs rather than in the front of their bodies since it's easier.

"When I went home for my first visit after moving to America, my mother and friends said they noticed a change in my posture."

"My mother would scold, saying: 'You are not walking or sitting straight. It is because of the cars!' "

Janet gained quite a bit of weight in America because she walked less. When I asked if she held her stomach in tightly (the common concept in our culture for good posture) she seemed shocked. "Why would I?" I explained that it represented good posture in the U.S., and that holding the pelvis in a position that contracted the abdominals and buttocks was thought to "protect the back."

"No, we do not," she quickly replied, "I thought you just knew naturally how to move. No one taught us how, we just did as our parents and grandparents did."

Unfortunately, so do we, but our role models have unwittingly led us astray.

It was amusing to her when she found out that Americans worked on the way they walked or their posture.

She learned to carry items on her head when she was eight. This was part of a ceremony where she carried a clay pot on her head, symbolizing a young girl coming into her womanhood. When she was learning, her grandmother told her to "let the pot be a part of you" which made it easy to carry.

It is customary for the women of the village to gather firewood. They might spend a day traveling to find a tree to cut down, then walk miles home with up to 50 pieces on their head.

"Women do all the washing in the lake, a five-minute walk from where they live. It's normal for them to bend over for long periods of time doing washing, then to carry water home. As the women get older they don't stop doing the chores. My grandmother's 80 years old and still does her share."

Janet said that in their culture, people did not complain about what life had dealt them as they believed that all things, including movement, had meaning. She told me she never recalled her people complaining of back or joint pain.

Mary's Story

Mary grew up in the Philippines. When she and Janet met in the United States, they found to their surprise they shared almost identical lifestyles and traditions despite being on opposite sides of the globe. Their movement, their way of cooking and certain traditions within the family were the same. Mary's family members, however, carried baskets of fish on their heads instead of clay pots of water.

How we move is a cultural thing. What we consider normal posture has been skewed by watching movie stars, models and advertisements that emphasize our sedentary lifestyle. It does not encourage us to move with skill or awareness, although the body was created to follow certain principles of movement.

Lincoln is fit. He's an excellent swimmer and has helped lead whitewater kayaking tours in Nepal. When you think of Nepal, your next thought is probably "Himalaya Mountains." The Kingdom of Nepal is a small country that rises from the plains of the River Ganges in the south to more than 29,000 ft. at Mt. Everest, all within 150 miles. The people are small and wiry and, as Lincoln soon found out, can move loads with amazing strength and ease.

Local carriers are important help with the kayaking tours since you get from place to place by foot on trails. The hike for Lincoln began at 6,000 ft. and ended high at the Annapurna base camp at 14,000 ft., an elevation change that makes for intensive exercise.

"Thinking I was pretty fit, I decided to carry my own gear. I headed out but before I recognized what was going on, someone in sandals passed me carrying four or five cases of Coca-Cola on his back! I kept struggling up those very steep slopes, occasionally stopping for a break, thinking the whole time that I needed to set that pack down!

"When the carriers, including Mr. Coke, rested, they would smile and nod as I struggled past. 'OK,' I thought, 'they're taking a break, now is my chance to get ahead.' I tried to walk fast but before I knew it, the man with a huge weight supported by a tumpline across his forehead passed me.

Lincoln's Story

"Pretty soon I would catch up with him as he rested and he would smile and nod his head again. But basically, they walked me into the ground. What's more, I wouldn't even think about picking up the weight they carried on their backs, let alone carrying it any distance. These people kept the pace for seven to ten days at a time," Lincoln said.

He estimated that the carriers were in their 20s and 30s. People in Nepal do not have a long life expectancy, but regardless of their age, they continue to move with the same naturally upright posture.

Lincoln was so impressed by their carrying skills that he started taking photos.

"As I was looking at these people to see how they did it, I noticed they had very upright postures. My first thought was they were all in the military because they were standing up so straight! Then I realized that all the people seemed to have very straight backs and upright posture all the time. It's even obvious when they sit down to talk with you. It kind of catches you off guard because

"I realized that all the people seemed to have very straight backs and upright posture all the time."

they are sitting so straight but they will be looking at you with a relaxed expression on their face. They don't have tension in their faces like people in military training. "These people are very lean and extremely strong and I've worked with them where we are lifting things and loading things onto rafts and onto buses. It's frightening how strong they are.

"My perspective now of movement, like walking, is that for westerners it's almost a chore. If we were looking at a Nepali, he would cover the same distance in half the time and the movement would be flowing."

Historical Perspective

Is our way of movement today the way we have always moved? The past century saw an unprecedented amount of change. Everything from where we live and how we live, to what we eat and how we get it, to how we travel and communicate. Even what we value has changed enormously.

We moved from an agricultural society to become one of the most urbanized countries in the world. We moved into mass production, into the belief that time-saving and productivity were all, and into the age of advertising. How have these factors affected our bodies and posture?

The last century's focus of style and dress began its big shift in the early 1900s. Fashions and styles, especially for women, usually reflect the social mores of the times. The corset and tight shoes symbolized the restraint of the Victorian era, and women started throwing these away to symbolize their independence.[11] In 1912, one woman was even arrested for going out in public without a corset. The restrictions of the corsets and shoes might have been bad for breathing and feet but it was nothing to what would happen in the "flapper" era. That's when the tucked buttocks arrived to stay.

The Roaring '20s was more than dance, music, parties and daring. Show business boomed and advertising became respectable and highly lucrative. With "image" paramount, what about posture?

The role of posture became cosmetic for both men and women and the purpose of posture began to leave. Even though the Victorian dress was exaggerated with its corsets and other adornments, the style was similar to women working on the land with emphasis on a strong, straight spine that allowed them to move freely from the hips. Now the desired look was to drape your body over chairs and couches.

The auto industry created a new frontier, and lifestyles became more sedentary both on the factory floor where the vehicles were manufactured and for the drivers. Life was now focused on time, on getting it done instead of enjoying doing it.

In 1914, the first moving assembly line in automobile history appeared at the Ford factory in Detroit. As Henry Ford

> The role of posture became cosmetic for both men and women and the purpose . . . began to leave.

said himself: "Every piece in the shop moves. No workman has anything to do with moving or lifting anything. Save 10 steps a day for 12,000 employees and you will have saved 50 miles of wasted motion and misspent energy."[12]

We were moving less. This was even reflected on the farm where machines were taking over.

Advertising boomed, and we all learned the image to adopt. President Coolidge himself acknowledged the image of the country was being manipulated when he said, "It [advertising] is the most potent influence in adopting and changing the habits and modes of life, affecting what we eat, what we wear, and the work and play of the whole nation."[13] He didn't realize it would even affect the most fundamental part of us: the way we move—our posture.

Can clothes affect the way we think? When actress Shirley Maclaine addressed graduates of the Fashion Institute of Technology, she asked them not to design only tight clothes because your soul needs room to move: "The looser the clothes, the looser the skin, the wiser you will become."[14]

Scientific Back Up

So other cultures, and our forebears, move differently from how we do today. What have scientists thought about this? When the spine loses its natural range of motion, especially in the lumbar area, it influences how all joints move.

Bantu children move as their parents do and lose very little lumbar (low back) spine range of motion as they age, according to a very interesting comparative study by Jonck and van Niekerk.[15] They found that reduction in mobility of the lumbar spine when bending forward and backwards (flexion and extension) in Bantu people occurred in a gradual and regular fashion. In contrast, Europeans showed a more abrupt reduction in mobility between the ages of 13 to 35. The lowest segments of the spine, where there is the greatest mobility, were measured and by the age of 34, Europeans had only 8-12 degrees of motion in their spines. By contrast, Bantus still had 19 to 21 degrees of motion in forward and backwards bending between the ages of 30 and 40. Jonck and van Niekerk stated that they investigated and studied the motion of the lumbar spine of Bantu people for two reasons: "Presumed greater stability of the Bantu lumbar spine as compared with the European and in view of the rarity of the disc syndrome in the Bantu."[16] Western posture changed to accommodate the styles and ways of the environment and this change was exacerbated when we taught good posture that was counter to the natural posture of children. We have lost good spinal architecture. Other cultures appear to have worked with gravity when making decisions on how to move, bend or lift things. We adapt, almost unknowingly, to match what we see around us, thus eliminating the conscious awareness of gravity at work.

Other cultures appear to have worked with gravity when making decisions on how to move, bend or lift things.

The Journey Recapped

Body Balance gives us the principles or guidelines necessary to restore natural movement, that is, to regain Universal Movements. Universal Movements are the simplest elements of movement necessary for all movement. Once we have these simple moves correct, we use the same principles for our dancing, tennis and other activities. Four steps brought me to a very simple answer to many complicated problems. (1) I trained in physical therapy to help people heal. But I saw how the clinical world and reality diverged. (2) I broadened my views on therapy through yoga, "yoking" the concepts of integration and simplicity of movement with the influences of gravity as a vital force. (3) I used Angie Thusius's guidelines and a global perspective of movement as a foundation on which to build a methodology to restore our natural movement we had as children. (4) I started searching for the scientific basis of why these principles I saw healing people—the *Body Balance Method*—actually worked.

The key to *Body Balance* is integration. We must learn how to integrate education into the real world, to integrate ideas from other cultures into our own, and to deal with our bodies as integrated wholes so we can be healthy individuals.

The boy's hips are tilted forward creating a natural arch, (look at the waist band). The boy carries more weight in the back of the body through his upright thigh bones.

The father's hips are tilted in the opposite direction (look at the seat of his pants). His shirt would hang straight if he were standing correctly.

Get creative

Get creative about what kind of movement you would like to do. Don't think about calories, getting fit, or what you SHOULD do but what you want to do. If you are already active, ask yourself if it is something you really love. Maybe there are other sports or activities that you would have more fun doing. Then . . . just do it . . . with the Body Balance principles as your guide.

Kinesthesia is the conscious awareness of joint position and movement or the ability to perceive extent, direction or weight of movement according to Garn & Newton as quoted in B. Brownstein & S. Bronner (eds.), *Functional Movement in Orthopaedic and Sports Physical Therapy: Evaluation, Treatment, & Outcomes*, Churchill Livingstone Inc. New York, 1997, p. 43.

1 Since this time, the process has changed and similar knee rehabilitation has become far more patient-friendly.

2 B. Brownstein and S. Bronner (eds.), *Functional Movement in Orthopaedic & Sports Physical Therapy: Evaluation, Treatment & Outcomes*, ed. Churchill Livingstone Inc., New York, 1997, p. 141.

3 M. Csikszentmihalyi, *Flow*, Basic Books, New York, 1997, p. 21.

4 B. K. S. Iyengar, *Light on the Yoga Sutras of Patanjali*, Thorsons, California, 1993, p. 1.

5 Ibid.

6 B. K. S. Iyengar, *Light on the Yoga Sutras of Patanjali*, Thorsons, California, 1993, p. 19.

7 Silva, Mira, & Shyam Mehta, *Yoga the Iyengar Way*, Alfred A. Knopf, New York, 1990, p. 8.

8 Ibid.

9 B. K. S. Iyengar, *Light on the Yoga Sutras of Patanjali*, Thorsons, California, 1993, p.p. 28-29.

10 A. Thusius, J. Couch, "Living on the Axis," *Yoga Journal*, July/August 1991.

11 P. Humphrey (ed.), *America in the 20th Century*, Marshall Cavendish, New York, 1995, p. 35.

12 Ibid, p. 208.

13 Ibid, 1995, p. 325.

14 L. Beil, "Girth of a Nation: Doctors warn of health crisis as obesity gains on Americans," *The Dallas Morning News*, August 29, 1999, p. 1.

15 L. M. Jonck & J. M. van Niekerk, "A Roentgenological Study of Motion of the Lumbar Spine of the Bantu," *Journal of Laboratory & Clinical Medicine*, June 1961, p. 68.

16 Ibid.

CHAPTER III
THE LOOK

"The Masai men and women in Africa probably have the most elegant spines in the world ...Walking for miles across the expanse of plains, or standing amidst a herd of cattle, their tall, lean bodies are studies in ease and beauty. Their houses, in contrast, are low with tiny doors and few windows. To go inside, one has to bend low and maneuver carefully through small passageways. Once inside you remain folded toward the ground." [1]

CHAPTER III

THE LOOK THAT LETS YOU MOVE IN COMFORT

This chapter gets to the base of the differences between the "optimal posture" of *The McCall Body Balance* and generally accepted "good" posture. It asks you to observe the logic of the *McCall Method* and then shows you "the look" to look for in your body.

You don't have to gain the "optimal posture" to get your joints moving and feeling better, but you must start heading in the direction that will create healthier movement.

It's time to look at the body from a different perspective because existing ideas on posture and movement do not seem to be working. Here are three pieces of evidence:

1. 80 percent of all adults will experience back pain;
2. Back pain is the leading reason for visits to doctors' offices, surgery, hospitalization and work disability;
3. Back pain costs in the United States are $50 billion annually.[2]

There is no healthy, functioning model of movement in western societies. We continue to perfect how to replace our joints rather than prevent joint degeneration.

Richard Deyo, M.D., writing in *Scientific American*,[3] put it well: "Medicine has at best a limited understanding of the condition [low back pain]. In fact, medicine's reliance on outdated ideas may have actually contributed to the problem." Dr. Deyo, a general internist and a professor at the University of Washington, should know. He has a long-standing interest in clinical research on low-back problems.

You need to do more than increase your daily movements; you need to know how to move—how to sit, stand, bend, walk and turn. You need to rebuild your basic physical knowledge because chances are that you have lost the skill of movement.

The First Step in Scientific Methodology: Observation

More than 400 years ago, Frances Bacon presented what today is called "hypothesis-based research" for scientific rationale.[4] Observation is the first action in this type of research.

Nobel Prize-winning physicist Richard Feynman, in his book *The Meaning of It All*, states that one of the three meanings of science is "finding things out." This, he says, is the method based "on the principle that observation is the judge of whether something is so or not."[5]

He goes on to discuss the expression: "The exception tests the rule." As he says, this really means: "The exception proves that the rule is wrong."[6]

It is my observation that proper movement is the key to healthy tissue, but most people have no understanding of "proper movement." To regain this body wisdom you need to start by redefining optimal posture, which is not just a static state.

Optimal Posture

Optimal posture is much more than standing, sitting or moving well. A different and more useful way to define optimal posture is: standing, sitting or lying with as little muscle activity as possible, then moving correctly to create a natural transfer of weight. Think of it as going from one place to another with the least amount of effort.

Research consistently supports the theory that when standing or sitting, the least amount of work needed to maintain the upright position is the most ideal or optimal. For instance, whether you are stationary or moving, it should take the least amount of muscular activity to keep your bones in balance.[7] Further, the less work the neuromuscular system must do in static positions, the more ideal. In other words, the less your body has to work, the more the parts are in place.

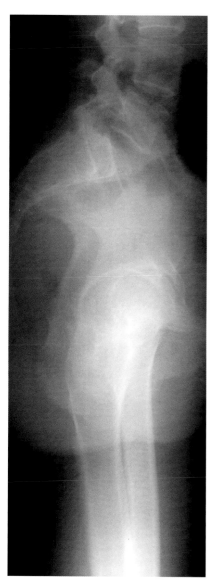

We continue to become more advanced in how to replace our joints rather than in how to keep them.

deSiGN Design Features

Gravity plays a vital role in keeping our body healthy. You suffer unnecessary wear and tear when you are out of line with gravity, which is what I believe happens because of today's view of "good" posture.

The first "design feature," according to *The McCall Body Balance Method* is that your body should be in a relaxed state.

Compare this with the prevailing theory of optimal posture. This view holds that when the pelvis is in the correct position, the buttocks should be kept in a slight posterior tilt (tucked buttocks). This is achieved by using the abdominals and muscles of the buttocks and pelvic floor when standing, bending or lifting. This theory does not stand up to the definition of "optimal." There is no need to voluntarily "hold" any body part when standing, moving or lifting. When our bones are stacked correctly all muscles are ready to work as needed. A natural anterior pelvic tilt occurs if the stomach and buttocks are relaxed.

As well as relaxation being the natural mode, our bodies must use gravity to its advantage. Gravity plays a vital role in keeping your body healthy, but suffers unnecessary wear and tear when it's out of line with gravity, which is what I believe happens because of today's view of "good" posture.

Gravity gives everything weight and resistance so you literally react to gravity by changing your body position while attempting to balance your structure around a particular point.[8] Your daily movements are a reaction to gravity since it constantly pulls you to the ground as you lift or move. This creates healthy stress, if all the forces balance out. This force stimulates bones to stay strong when you move in harmony with gravity but can contribute to breakdown when you're out of balance. When you stack your bones as they were designed with gravity in mind, your bodies work less and gain more strength.

How it all stacks up in your body:

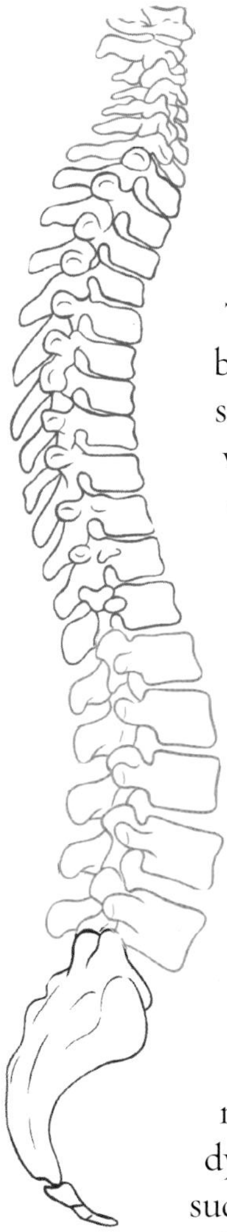

cervical
(neck)

thoracic
(chest)

lumbar
(low back)

Your spine can be divided into four sections. The lowest section is your sacrum; this attaches to your pelvis and to your lumbar spine, which goes up about as far as your waist. The middle section is your thoracic spine and rib area which contributes the most to the length of your spine.[9] The top area is your neck or cervical spine. The spine is a series of multiple vertebrae, each composed of body, with a smaller bony portion attached called the vertebral arch.[10] Sandwiched between each vertebral body is the intervertebral disc. When you run your fingers down your spine it's the spinous processes on these arches you feel. The vertebral bodies and intervertebral discs should carry most of your weight because the smaller vertebral arches break down if they are loaded with too much weight. Incorrect loading on the vertebral arches contributes to many common back dysfunctions today, such as osteoarthritis and degenerative joint disease.[11]

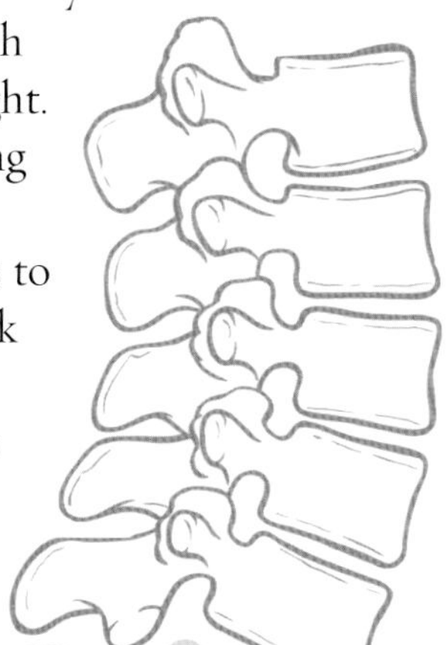

Your spine

Your
Head

Your cervical spine carries the least
amount of weight but its position relative to
your spine greatly influences how the rest of
your body works. There is general agreement
that your head, neck and upper spine should
be over your pelvis. This is because a structure will be strongest when the weight is
distributed evenly over the greatest surface
area. Again, your body and its joints will last
longer and work better if weight is spread
over the greatest surface area of any joint.
This is particularly necessary in the lower
spine and in the pelvis, which carries most
of the body's weight versus the upper
spine, which carries less of the load.

Forward head
position is a
common
postural
misalign-
ment.

Forward head position is a common postural misalignment. If the transitional area between your head and neck is incorrect, it can cause headaches.

Look at the picture of the women carrying pots on their heads. The pot balances on the women's heads which balances on their cervical spines. The pots on their heads give them sensory input to keep their heads in line with their spines. Since you don't carry things on your head, it's no surprise that the first place you lose your sense of alignment is in your head and neck. Think about how "forward" you are each day: over a computer, at a steering wheel or even in conversation.

These Native American women are examples of cultural influences on the way they use their necks.

Thoracic spine; the rib area

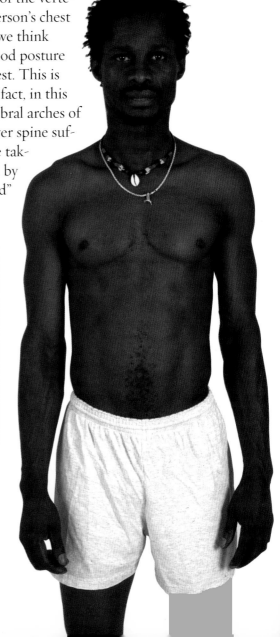

chest

In the thoracic spine or rib area we often put too much weight on the back of the spine (vertebral arch) instead of the vertebral bodies. If a person's chest sticks out and up we think that person has good posture as well as a big chest. This is very deceiving. In fact, in this position, the vertebral arches of the upper and lower spine suffer because they're taking too much load by this assumed "good" posture.

The bodybuilder represents a creation that is not built for function and is NOT relaxed. The man on the right developed his body by moving naturally all his life. He teaches West African dance and drumming. His chest, neck and body have normal natural development and he IS relaxed!

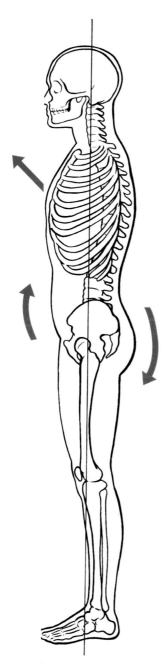

Even the slightest lifting of the chest can shift the weight off the body of the vertebra and onto the arch. In turn, this does not allow postural muscles in that area to work as they should because the spinal segments are no longer in their proper place. This can lead to instability in the areas above and below causing breakdown and pain. This means the low back and neck get the bad stress.

Even the slightest lifting of the chest can shift the weight off the body of the vertebra and onto the arch.

Traditional medical illustration of correct standing posture. Buttocks tucked, chest lifted.

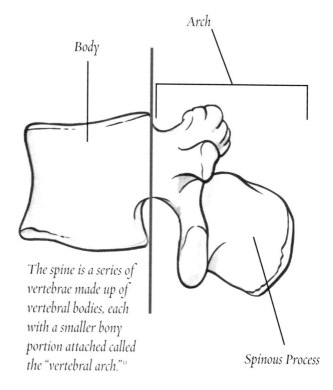

Body

Arch

The spine is a series of vertebrae made up of vertebral bodies, each with a smaller bony portion attached called the "vertebral arch."[11]

Spinous Process

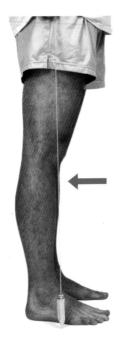

Locking the knees out will place the weight too far forward in the knees.

The buttocks are tucked using the abdominal and pelvic floor muscles. This non-relaxed posture puts weight too far forward in the knees. (Traditional "good posture")

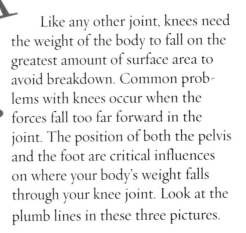

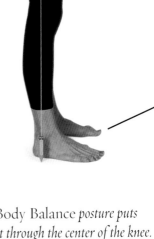

The Body Balance posture puts weight through the center of the knee.

K
n
e
e
s

Like any other joint, knees need the weight of the body to fall on the greatest amount of surface area to avoid breakdown. Common problems with knees occur when the forces fall too far forward in the joint. The position of both the pelvis and the foot are critical influences on where your body's weight falls through your knee joint. Look at the plumb lines in these three pictures.

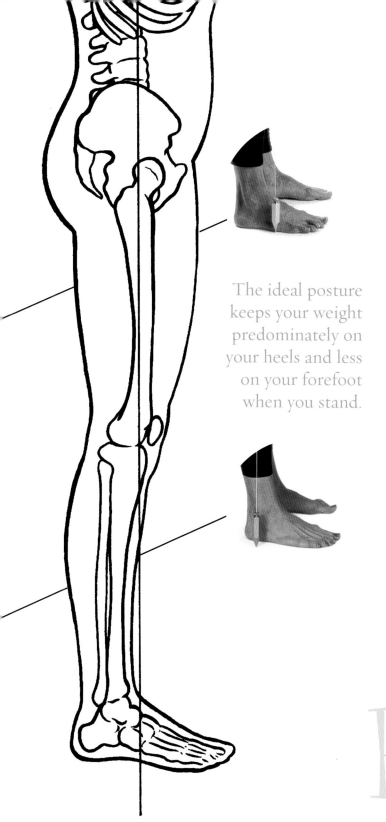

The ideal posture keeps your weight predominately on your heels and less on your forefoot when you stand.

Traditional medical illustration of "good posture"

If your weight is forward of your ideal center of gravity, then the line of gravity will fall through the arch of your foot. The ideal posture keeps your weight predominately on your heels and less on your forefoot when you stand. Everything is connected. The "functional strength" of the muscles in your foot depends greatly on how you use your foot, which is directly related to how you stand and walk. Having your bones stacked correctly throughout your body allows the forces to fall through your foot without creating tissue breakdown. It also allows correct transfer of weight throughout the foot when you walk. If there is more weight in your forefoot when you stand, the mechanics of your gait will change for the worse. Too much forefoot weight also alters the ability of your ankle and foot to react as needed. These weight imbalances ultimately influence your whole structure.

Feet

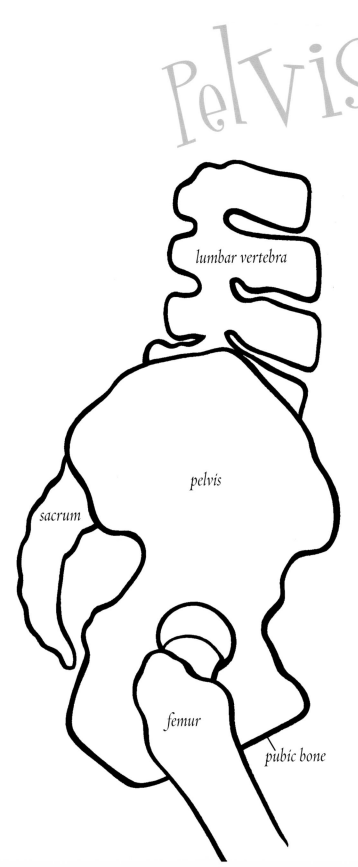

Pelvis

lumbar vertebra

pelvis

sacrum

femur

pubic bone

The expression "structure governs function" can help you understand that how your bones are arranged will govern how your muscles act. Before starting a movement, each joint has a unique setup position. If the structure is in the "ready" position, then fluid movement can occur. So the statement "structure governs function" applies to an individual joint or the body as a whole.

"Pelvis" is the Latin word for basin and, like a basin, the pelvis can tilt in many directions because of the way it is shaped and how it is attached to the spine and leg bones.[12]

When you are standing, how the bowl tilts will influence the spine and the legs, and is critical for your whole body. By tilting the bowl forward, the center of gravity shifts back. By tilting the bowl backward (as when your buttocks are tucked), the center of gravity shifts forward.

A fundamental principle of *McCall Body Balance* is that shifting the body's center of gravity back to the correct position provides a larger surface area for all the bones involved in weight-bearing, making it easier for your body to stand and move correctly. Probably the most significant difference between *Body Balance* and generally accepted theory begins in the pelvic area, where all movement begins.

Let's use the scientific principle of observation and compare some details of what has been considered for many years "good posture" and what *The McCall Body Balance Method* considers optimal posture.

A fundamental principle of *McCall Body Balance* is that shifting the body's center of gravity back to the correct position provides a larger surface area for all the bones involved in weight-bearing, making it easier for your body to stand correctly.

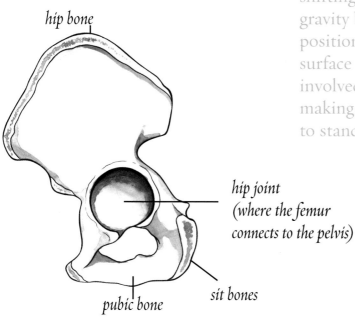

hip bone

hip joint
(where the femur
connects to the pelvis)

pubic bone

sit bones

"Pelvis" is the Latin word for basin and, like a basin, the pelvis can tilt in many directions because of the way it is shaped and how it is attached to the spine and leg bones.[12]

Body Balance

THE LINEUP

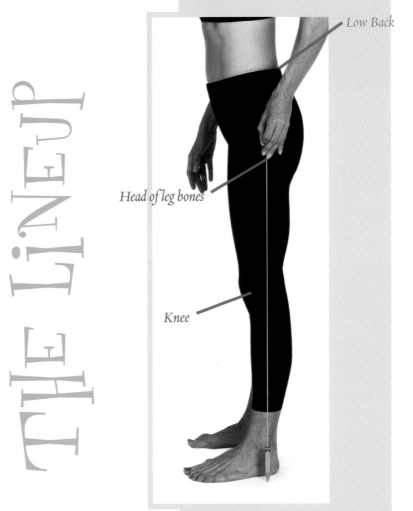

Low Back

Head of leg bones

Knee

The following series of pictures will show you the difference between *Body Balance* posture and other postures. The setup of these parts is critical to the correct alignment of the rest of the body.

Body Balance has more bones in line with gravity.

The Body Balance Posture

If you allow your stomach and buttocks to relax your pelvis will tilt forward and the leg bones will go back. More bone is in line with bone, therefore, you work less and build bone instead of breaking it down.

Popular Posture

Bad Body Balance

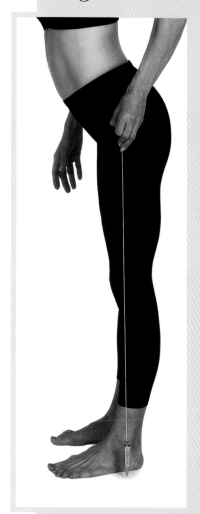

So-called "Good Posture"

If you tuck the buttocks, or
tilt the pelvis backwards,
the leg bones go forward.
This does not keep the leg
bones in line with the
spine, creating more work
to stand and move.

"Sway Back"

If you hike the buttocks
up in back, this moves
the lower spine forward
and out of line with the
leg bones.

Where the Weight SHOULD Be

This traditional medical drawing illustrates currently accepted "good posture." It clearly gives evidence how, if you stand this way, you in fact have less bone in line with gravity compared to the *Body Balance* posture.

──────── angle of pubic bone

● ● ● ● ● ● ● ● ● ● Centerline of head of femur

──────── Line of gravity (where the weight falls through the body)

 Sacrum

Lisa Ann standing in the traditional standing posture.

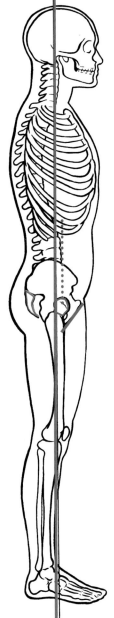

- See how the pubic bone is in front of the body.

- See how the sacrum is tucked.

- See how the head of the femur is not in line with the spine.

Lisa Ann in Body Balance
standing posture.

X-ray of Lisa Ann's pelvic area.

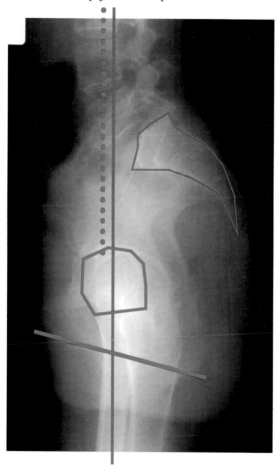

- See how my pubic bone is more horizontal or "under" my body.

- See how my sacrum is more behind my body, rather than tucked.

- See how the head of my leg bone is lined up with my lower spine.

X-samples

The X-ray on the left is of a 16-year-old female sprinter who complained of lower back pain. The problem is not the position of her pelvis or sacrum, but the position of her spine. It would be counterproductive to have her tuck her buttocks and tighten her stomach as in the traditional posture to correct her sway back. By doing this it could take away her ability to become a high-level athlete because she would lose her natural pelvic position, which is one reason why she runs so fast. Instead, her spine should be more in line with the head of her femur (as it is in my X-ray) to decrease her lower back pain.

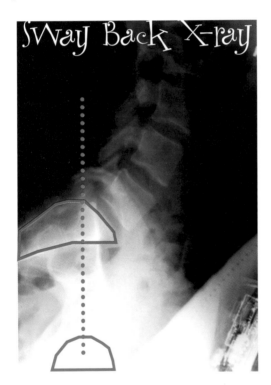

X-ray with "sway back" (misinterpretation of Lisa Ann's Body Balance posture). See how her sacrum is more behind her body, rather than tucked. BUT see how her spine does not line up with her leg bones.

 *Centerline of head of femur*

Sacrum

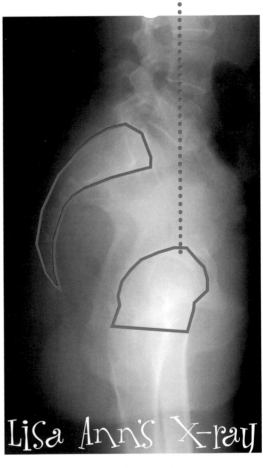

See how my sacrum is behind my body, rather than tucked. YET my leg bone is in line with my spine.

You Do The Test

Stand with your hands on the front of your thighs and tuck your buttocks so that your tailbone is down. Feel your legs come forward into your hands.

Then relax and lean slightly forward from your hips and feel your legs go back, (Make sure your knees are not locked.)

The pelvis must tilt forward to get the leg bones BACK, putting them more under your spine where they are meant to be. To learn to come upright correctly turn to chapter 4.

This drawing shows a much more natural shape of the torso when the body is in balance. Notice the abdominal area is not sucked in, the chest is not elevated, and the buttocks are not tucked under.

Putting It into Practice

sitting

Your spine and pelvis should be in the same position when you are sitting or standing; all that has changed is the supporting surface. The bonus is that now—with *Body Balance*—the transition between movements, such as between sitting and standing, will be easy because your joints are ready to move. Little effort is needed.

Your pubic bone is close to the horizontal plane, which makes it an ideal bone to sit on. When you sit in the center of the basin, you allow your spine to balance your weight and your muscles have the least amount of work.

As we sit and continue to stack our bones in their ideal placement, our sacrum slants forward and down in front just as it does when we stand. This is due to the forward tilting of the bowl which creates a natural arch in our lower spine.

The lowest vertebra is shaped uniquely for this natural arch to exist. That vertebra is wider in front than in back, fitting well as a transitional vertebra. This balances the spine, and assists the natural curves of the spine to maintain a more vertical position. The pelvis "carries" the trunk and makes it easier to eliminate poor posture seen in the upper spine, such as forward head and rounded shoulders. All parts of our structure are interdependent.

Your pubic bone is close to the horizontal plane, which makes it an ideal bone to sit on. When you sit in the center of the basin (remember "pelvis" is the Latin word for basin), you allow your spine to balance your weight, allowing your muscles to perform a minimum amount of work.

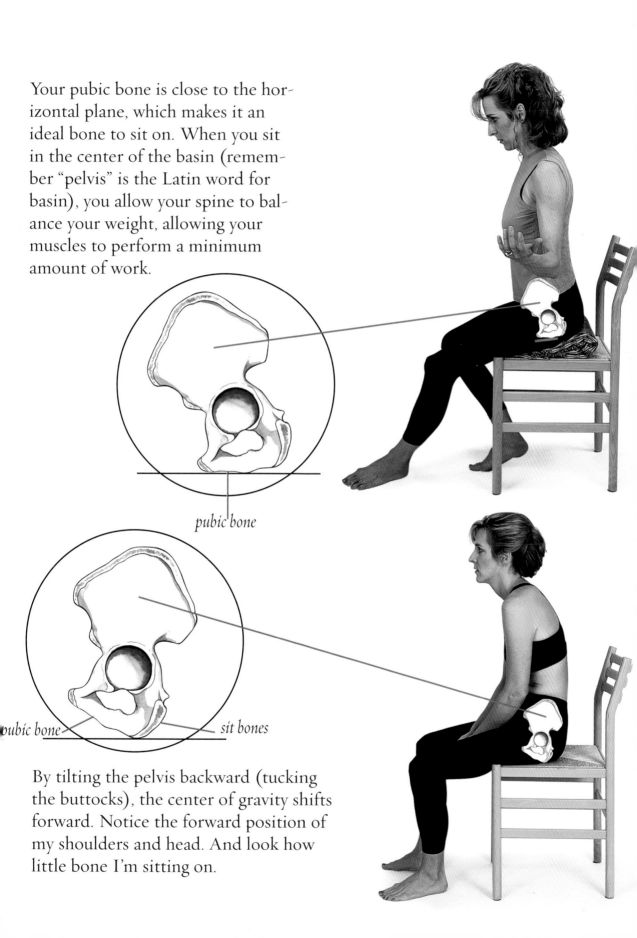

pubic bone

pubic bone   *sit bones*

By tilting the pelvis backward (tucking the buttocks), the center of gravity shifts forward. Notice the forward position of my shoulders and head. And look how little bone I'm sitting on.

BeNDing

The correct ways to bend and lift are much debated. Why don't people in less developed countries, who spend so much time bending, have the same level of back problems as people have in the West? It gets back to posture.

All joints respond to movement in the same manner. When joints work at extreme ranges, they are less capable of resisting tension. Joints work better when functioning in the midrange. You are more capable of doing work if you do not work your body with your joints locked or fully bent.[13]

What causes breakdown? Repetition of incorrect movement. If you repeat any incorrect movement often, you may create an injury. Fortunately, the reverse is also true.

In contrast, when you bend deeply at the hips to lift, you allow the forces to be distributed throughout the back,

The body was designed to lift over and over without breakdown. If you bend in a relaxed and natural way, there will be changes, good ones over time, that will increase your lifting skill. Experience it and believe.

Body Balance *uses the same mechanical principles as the crane. The cantilevered arm of the crane is similar to your back, straight and counterbalanced by weight (strength) in the buttocks and back.*

& Lifting

Bending as the body is designed to bend

Common lifting postures taught in our society today as "correct" may seem safe but it is not the way the body was designed to lift.

buttocks and legs. All joints remain in safe range of movement. People all over the world use this lifting technique daily. This way of bending allows you to safely use your back. If you do not use it, you lose it. So the back does not need to be protected by braces or moves that avoid its use, it needs to be used as it was designed to be used.

By relearning to bend and lift in this correct way, you start activating very powerful muscles along your spine in their natural manner.

Think of the logic of it by comparing this method with heavy machinery designed to lift weight. *Body Balance* uses the same mechanical principles as the crane. The cantilevered arm of the crane is similar to your back, straight and counterbalanced by weight (strength) in the buttocks, back and legs.

Most people lift, bend and move in whatever way is easiest at the time, which is a reflection of their posture and lack of understanding of how to move.

If the joints are not accustomed to moving in the midrange then bending naturally will need to be relearned.

These men bend at the hips to pull a large load. Notice the even groove down the spine. This represents an even distribution of forces working through the back to carry the load.

Putting It Together
In an ideal posture:

* Your back will have an even groove from the top of your shoulder blades to the base of your spine, illustrating the even distribution of weight throughout the spine. This is due to the development of the back muscles.[14]
* Your chest is broad and open, which means the shoulder blades have the appropriate placement on the back.
* Your rib cage looks broader at the top and tapers at the base.
* Your arms hang directly at your sides with your palms facing toward the body.
* Your spine looks straight except at the very base where there is a distinct angle that slants back and down. This is the lumbar-sacral junction.
* The front of your body looks long and, when viewed from the side, has a crescent shape with a continuous smooth line.
* Your upper thighs are rotated out.
* Your feet are only slightly turned out.
* Your kneecap is approximately in line with your 3rd toe.
* Your buttocks are not tucked or lifted but are literally "behind" your spine.
* Your head sits directly over your shoulders.
* Your face is relaxed, especially around your jaw area.

Notice how this Portuguese fisherman's arms hang directly by his side. His shoulders are squared, not slumped forward. His back gives evidence to a lifetime of healthy movement.

Buttocks not tucked but literally "behind" the spine.

The Look

We build a visual appreciation for many things, from what constitutes good art to what makes a prize-winning bull. Coaches spend years watching their athletes move and develop a trained eye for optimal movement in their sport. There are vast amounts of information to be gathered by observation, and the illustrations throughout this book are to help you see "the look" and how it applies to many different body types. Look at each shape, line and muscular development to better understand how to move. The pictures will help you develop an eye for this more ideal body.

The girl carrying the sticks smiles with all that weight on her head! Notice that her hips are back and her belly is relaxed as well.

IDEAL POSTURE

Child with same posture as young girl above. Notice how her head is directly in line with her shoulder and right leg. She stands stable, balanced and relaxed.

The Universal Movements

Universal Movements are the basic moves we make daily. From the perspective of pure mechanics, bones undergo only two types of movements: glide (translation) and rotation.[15] Everything visible on earth moves by these two mechanics. As a leaf falls from the tree, it moves through translation and rotation. A movement may look more complex, but the only difference is the axis on which the movement occurs. For example, if you bend forward at your hips you are rotating about the "x" axis. If you twist you're moving on the "y" axis.[16] The simplest movement, such as opening a car door is a much more integrated movement than you may realize with rotation on a three-dimensional axis. As you learn to move better, you can regain more balance at the joint level.

y-axis

x-axis

Improving What We Have

Not everyone can regain an ideal alignment. Many people have structural or joint uniqueness and react to ground and trunk forces in unique ways. Changes due to injury, surgery or age may also prevent the ideal tilt of the pelvis, groove down the back or perfect rib positioning. However, everyone has the ability to improve. I'm frequently amazed at the changes people do achieve. The key is putting these principles to work.

Small changes make a huge impact. These changes create a shift in the center of gravity, which eliminates many of the unwanted forces that act on your body as a whole. When the architecture of your bones is allowed to operate as designed, or close to it, your neuromuscular system, which gives you the feedback on how to move, is allowed to work effectively, and can create fluid movement.

Small changes make a huge impact.

Where to Start

The good news is that there's growing interest in how your posture influences your health. The medical community, sports teams and local gyms are using terms such as balance, alignment, functional exercise and sport specific training, all of which are about how to move.

The road to the optimal moving posture does not have to be as complex as many people tend to make it. It doesn't need a multiplicity of exercise and stretch routines. The answer is simple. To get back to the basics of how to move without pain, you need to use what you already have. You are a well-designed, mental and physical being who can care for yourself. If given the right information, those moves that move you daily can bring you the optimal answers.

L○○K

When watching TV, read-
ing a magazine or glancing
at a billboard notice if the
posture or "look" is a reflection of comfort or is trying
to sell you a product or idea
(and their posture).

1 Andrea Olsen, *BodyStories*, Station Hill Press,
  New York, 1991, p. 51.
2 'Low-Back Pain,' Richard A. Deyo, *Scientific American*,
  August 1998, p. 3.
3 Ibid., p. 3.
4 R. Brand, "Hypothesis-Based Research," *The Journal of Orthopaedic
  & Sports Physical Therapy*, August 1998, Vol. 28, No. 2, p. 71/72.
5 R. P. Feynman, *The Meaning of It All*, Perseus Books, Massachusetts,
  1998, p. 15.
6 Ibid, p. 15.
7 B. Brownstein and S. Bronner (eds.), *Functional Movement in
  Orthopaedic & Sports Physical Therapy: Evaluation, Treatment & Outcomes*,
  ed., Churchill Livingstone Inc., New York, 1997, p. 149.
8 Bryan Williamson, P.T., MS, interviewed by Lisa Ann McCall,
  August 1999, Dallas, Texas, personal.

9 L. T. Twomey and J. R. Taylor, *Physical Therapy of the Low Back*,
  Churchill Livingstone, New York, 1994, p. 66.
10 N. Bogduk and L. Twomey, *Clinical Anatomy of the Lumbar Spine*,
  Churchill Livingstone, Melbourne, 1991, pp. 7, 66.
11 Ibid., p. 149.
12 Williamson, op.cit.
13 Ibid.
14 A. Thusius, J. Couch, "Living on the Axis," *Yoga Journal*,
  July/August 1991, p. 71.
15 N. Bogduk and L. Twomey, *Clinical Anatomy of the Lumbar Spine*,
  Churchill Livingstone, Melbourne, 1991, p. 53.;
  BW Interview.
16 C. Z. Hinkle, *Fundamentals of Anatomy of Movement*,
  Mosby, Missouri, 1997, p.p. 7, 82.

CHAPTER IV
MOVES THAT
GET YOU MOVING

CHAPTER IV

MOVes That Get you MOVIng

This chapter is the heart of the book. It tells you how to move. All daily movements are the building blocks of more complex action, and a movement routine should be designed with that in mind. All the moves I teach are to build a pattern of understanding in the neuromuscular system. This means that by repeating a movement, it becomes part of you, a reflex action.

The Road Rules
When in doubt, read the directions

Rule 1. Make the old moves Work better

For instance, moving your body from standing to lying may seem unimportant, but when this move is broken down into its fundamental parts, it re-educates your body so you will move and feel better. *Body Balance* is a refresher course for your joints on the fundamentals of movement.

Rule 2. Make it easy on yourself

Start simply with *Body Balance* moves. Integrate one to three movement principles so you can see changes. For example, you may want to start by sitting with an upper back support in your car, by standing in a more balanced relaxed way or by using *Body Balance* walking. Do these for a week or two and then add or switch your focus to another one or two moves.

Rule 3. Position your pelvis

Your pelvis should be in the correct position for the desired movement. All movements have a starting point. When you begin the moves at the pelvis correctly, they will be easier and more powerful. The flow-on effect influences all the other joints involved. For example, athletes such as golfers, skaters and hockey players must get their pelvises in position to move correctly.

Rule 4. Think bones

Think about moving your BONE; do not think about muscles. If the instructions refer to say, the inner shoulder blades, picture that place in your mind, thinking of the relevant bones.

Rule 5. Relax and integrate

The secret to making big changes is to relax and integrate. These moves will begin to stick and the body will naturally move correctly when these principles are integrated into your life. Relax all other areas that are not involved in the moves, unless instructed otherwise. It may be difficult to relax because these areas have not responded to your commands in years. What's a relaxed movement? Move your arm back and forth from the elbow a few times. You are working muscles in the back and front of the upper arm, but it does not feel difficult or like work. It feels easy. Bending at the hips can feel that easy over time by using *Body Balance* principles.

Rule 6. Don't force the issue on tissue

Injuries occur if you use too much force, so focus on technique. Doing the movement correctly is what counts—just like when you're learning to swing a golf club or to dance. Forcing the body into position can irritate tissue. If you want to make the correct changes and learn more quickly, repeat the moves over and over in your daily life. You'll also feel more flexible and stronger. Feel, don't force, movement. You are learning a skill, and it's like learning archery, fencing, target shooting or ping pong. Once you get the feel, your moves will be stronger with less effort. You are learning to feel movement at the joint level.

Rule 7. Time your breaths

Each move will have specific guidelines that focus on breathing to increase relaxation. For example, instead of counting five counts, you will count for three breath cycles. These are powerful moves because you are working muscles at the joint level, which is not what you are used to doing, so a few repetitions can wear you out. Integrate the concept into daily life to create the repetition you need.

Rule 8. Sit around and do nothing

Incorporate a static or passive resistive pose soon. I like people to integrate a static "exercise," such as sitting reclined, frog or side lying, into their lives. These positions encourage relaxation, and changes will occur without your help. They also help you gain the feel of the more active movements. You will be impressed at how much change occurs while doing NOTHING but allowing gravity to work on you.

Rule 9. Overlap the moves

All the moves build on each other. What you learn on the floor will be used in standing and sitting. You will gain a new awareness or feel at the joint level that will make all movements flow with the innate knowledge. There are not too many different moves; they just fit into different forms depending on whether you are on your buttocks, on your back or walking down the street.

Rule 10. Advance and relax

An advanced Body Balancer can relax and do the movement better; it's almost second nature because you have practiced it and need less mental or physical effort. You will be able to relate to what top athletes say, that their moves are so easy when they are tuned into the feel.

Rule 11. Don't mix and match

Don't mix these moves with other stretches. If you are under the care of a health professional, finish his or her instructions before beginning these moves. This work does not mix well with yoga or other forms of movement therapy. Your body is trying to learn a new skill to integrate in your workout routine and daily activities.

Rule 12. Feeling no pain is gain

You may feel an ache, a little warmth, stretching, slight pulling and soreness, but if you feel pain, stop what you are doing. You are unlocking the secret to feeling moves close to your joints. This should not create pain. Relax and try the move again with a little less effort and a lot more awareness.

INDex oF the MoVemeNts

B Beginner

I Intermediate

A Advanced

Ribs	Shoulders
Standing	Lying
Bending	Neck
Feet	Balancing
Tricks	Floor
Sitting	

The floor movements are the building blocks that make all the other moves have a feel and create the understanding of a new awareness. Movement is a form of language.

1.1 Chest Hang B

Life Goal:
To learn how to support your back when sitting in a chair comfortably.

Body Goal:
To regain natural shoulder movement.

Props:
Two pillows and a couple of blankets or a quilt.

Method:

Lie supine on the floor, then prop yourself up onto your forearms.

Walk your forearms back towards the pillow with your elbows pointing away from your body. Allow your chest to drop at mid-back. Gently press upper and mid-back toward the floor, feeling a stretch in the shoulder blades, mid-back and in the front of the shoulders. Keep your head forward.

Do not allow your shoulders to ride up toward your ears.

Feel your back broaden as you slowly slide your elbows out to the sides and go to the floor. Use your hands to place your neck on the pillow by gently pulling your hair at the nape of the neck so the back of the neck will straighten. Bring your shoulders down away from your ears.

The pillow should be positioned to create a wedge beginning at the top of your shoulder blades.

1.2 Rock B

Body Goal:

To relax muscles in the lower body. This is a simple repetitive movement to get the back muscles to relax and to get the synovial fluid (the oil for your joints) moving in the back. Do this in between the other floor moves or any time you feel a little stiff in the back.

Method:

Lie supine on the floor. Bend your legs, keep your feet flat on the floor, heels approximately 15 inches apart.

Rock both your knees side to side repeatedly for 2 to 5 minutes.

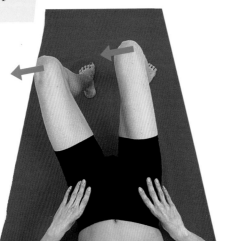

1.2.1 Little Roll   B

Body Goal:

To help regain hip movement for proper bending, standing and sitting. It strengthens muscles deep in the buttocks used for locomotion and "opens" the front of the hips, allowing the trunk muscles to work more efficiently.

Life Goal:

To increase comfort in standing, bending and walking and sitting.

Method:

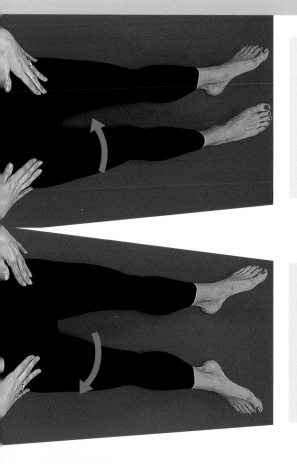

Begin on the floor, supine. Relax all muscles and place your hands on your hip bones in front.

Start rolling your right leg inward, then slowly roll your leg bone out as far as possible feeling the movement from high up in the hip (focus on moving bone). Keep your thigh rotated out as you relax everything else.

Now focus on your heel and visualize someone pulling your leg gently down and away from your hip by the heel. You will feel as if you're lengthening your leg while keeping your foot, knee and thigh relaxed. Hold this position for three breath cycles, focusing on relaxing all other areas of the hip, back and body.

Relax, and do your left leg.

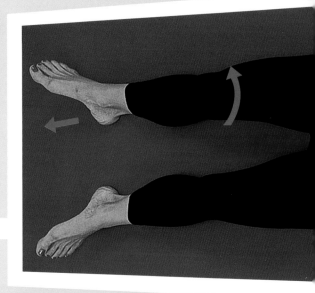

Repeat
3 to 4 times on each leg, alternating legs. Closing your eyes will give you an awareness advantage. (These small movements are very effective.)

1.22 Big Roll   B

Pelvic Position:

Keep the supine position. The lower back will have a soft arch, which means the pelvis will be tilted slightly forward. Do not flatten the lower back by pressing it to the floor. If the lower back feels a little tight, lift the buttocks, lengthen the back out and then rest it back on the floor, re-relaxing all muscles.

Body Goal:

To get results similar to the Little Roll but affecting a broader area of the hip.

Life Goal:

To stand on one leg comfortably and more balanced; it aids walking and bending and sitting.

Method:

Place a pillow a few inches from your right knee. Lie supine on the floor. Bend both legs, feet resting on the floor.

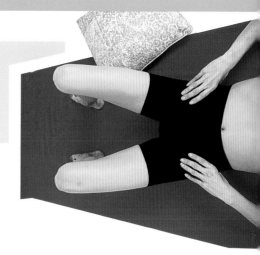

SLOWLY and ACTIVELY open your right leg out onto the pillow next to your knee. Place your finger-tips on your hip bone to gain the opening sensation at the hip as you actively rotate your leg onto the pillow. Keep the muscles active deep in your hip even though your leg is resting on the pillow.

Note: The foot of the leg that is bent will be close to the calf muscle on the opposite leg.

Repeat

Bend both knees again, place the pillow on your left side and repeat movement for your left leg.

Slide the other leg down, turn it in, then out doing a little roll. Do not grip your buttocks muscles. Don't forget to breathe and relax.

1.3 Rib Tilt B

Life Goal:

To get rid of poor spine posture including forward head, rounded shoulders and sway back.

Body Goal:

To increase abdominal or trunk strength. It helps bring the upper spine and neck in line with the rest of the spine while creating a stronger trunk (both abdominal and back) muscles. This interconnects with all movements.

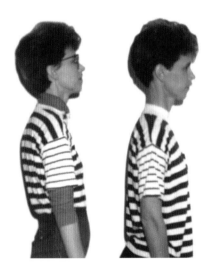

An example of the effects of better spine posture.

Method:

Repeat

Do this workout 2 or 3 times. The second time, close your eyes and visualize the move. You will feel it more. This entire workout on the floor can take 10 to 15 minutes. It is a good way to relax at the end of the day or to start the day. It gets rid of stiffness and brings a new awareness of joint movement.

Focus on a point on your back in line with the base of your breastbone. Press this point to the floor. Stack your hands, little finger to thumb, on the back of your neck. Keep your shoulders down, away from your ears.

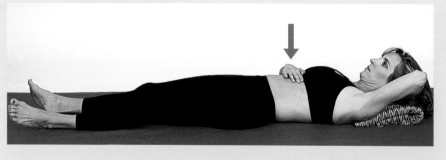

Begin to lift your head as you lengthen the back of the neck with your hands (keeping the face relaxed), making a large arc with your upper spine. Keep your focus on lengthening your spine as you come up only to the fixed point in your back that you are pressing into the floor.

Lower your head slowly, continuing the traction as you move toward the floor.

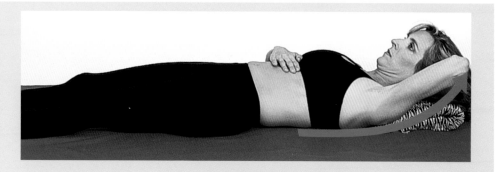

As you lower your head, keep one hand on your neck and the other hand at the base of your breastbone to keep the ribs from coming up. Your abdominals will be working.

Whether it's in a grocery line, an elevator ride, or even at a cocktail party, the time is right for a 30-second makeover that will help you stand longer than others. Practice this a few times in front of a mirror to get the hang of it and to realize it does not look as strange as it feels; then adapt the moves so your body will stand with ease for a long time. Take advantage of standing in line. Instead of being frustrated by the wait, you can make changes that will last a lifetime.

2.1 Stand in line to get the spine in line i

Body Goal:

To relearn to stand with more bones in line with gravity. The changing process will strengthen muscles deep in the hip, abdominals, the upper back and neck, and muscles in the feet.

Life Goal:

To stand relaxed with little effort, building a stronger, easier walk or run.

What to feel

Most people feel like apes when they first relearn to stand in balance. This will change the more you do the move. Most of the difficult work will be in keeping your chest from elevating and your lower back from over-arching as you come to the upright position. Focus on relaxing your lower back area to avoid over-arching.

Place heels approximate-
ly 7 inches apart with
your toes turned out
about an inch.

Place your fingers at the
tops of your thighs.
Relax your abdominals,
buttocks and pelvic floor
as you soften (bend)
your knees.

Relax:

Inhale and exhale,
relax the shoulders,
drop your arms and
release the rotation
of your thighs.

Shift most of your
weight back towards
your heels by guiding the
tops of your thighs back
with your fingers. Keep
your spine straight and

bend from the hips like
a waiter bowing. With
both big toes planted,
and knees still bent, turn
your knees out so your
inner knee faces forward.

Straighten your legs but
keep your knees turned
out and toes planted.

As you begin to bring
the torso up, focus on
a line between your
shoulder blades and
neck and lift up from
that line. As your spine
straightens keep your
chest from lifting. Keep
your big toes firmly
placed on the floor and
the weight in the heels
as you come upright.

2.2 Standing Quick Check ⓘ

Simplified:

Take advantage of those seconds in the elevator. Think weight shift and relaxation.

Pelvic Position:

The pelvis is tilted slightly forward with a soft arch in the lower back. The abdominals will feel stretched or lengthened and the buttocks are behind, not lifted or tucked.

Place your fingers at the tops of your thighs. With your knees bent but relaxed, shift your buttocks behind you with a slight forward tilt of the torso.

Then actively think of lifting the nape of your neck up while keeping your chest from elevating as you come upright. Do not let your hips move forward.

Ribs   Shoulders

anding   Lying

Floor   Neck

Feet   Balancing

Tricks   Bending

Sitting

3. Butt Up the Wall   B

Body Goal:

Build a better back, buttocks and legs. This is the move that brings strength and shape to more muscles than you thought you had. The back of your legs may feel like they are stretching, but they are really getting stronger. The buttocks and muscles of the back are getting a workout as well. The more you do this, the better built you will be.

Life Goal:

To learn how to bend. It's a move you do hundreds of times a day, from opening a car door to washing your face.

Props:

A wall and support for your arms, such as the back of a chair or a high ledge.

Method:

Stand with heels spaced approximately shoulder width apart and toes turned out one inch. Place the fingertips of one hand in the groove in the natural arch of the lower spine and the forearm of the other arm on a chair or high ledge in front of you for support.

Note:

Always start with support for the front of the body; otherwise, many muscles that must relearn to relax will not get the message.

Keeping your toes planted, bend your knees and turn your thighs out as you relax your groin and pelvic floor. Begin to bend at the hips by moving the sit bones of your buttocks back and up so they touch the wall. Maintain the groove in the back. If you feel your spine sticking out under your fingers you will not be bending at the hips but at the waist, so straighten your spine and the groove will come back (you may need to bend your knees more).

With the sit bones on the wall, rest your head on your arm, keeping your spine straight. Allow your body weight to shift forward toward the front of your feet.

Note:

If you tire, come up. Do not force yourself to stay in position, but with practice, it will be easier to stay longer.

Pelvic position:

The pelvis begins in the same position as in the standing posture.

Focus on pressing your sit bones into the wall as you relax the pelvic region; this is important for balance. Now relax into this position and breathe in and out for two breath cycles.

Coming out of Butt Up the Wall, draw your head down toward your chest and imagine you are coming up from the back of the neck. This prevents you from over-arching your spine. Keep your buttocks against the wall and allow them to slide down naturally.

The chair Janet sits in is from her homeland. It is made from two pieces of wood that hook together to create a chair. Sitting correctly is key otherwise the chair will collapse. The body, not the chair, must be ergonomically correct.

Sit Around and Get in Balance

Since we sit so much, we may as well be good at it. Sitting can cause profound changes, both good and bad, without their being noticed. Don't think of sitting correctly as being good for you, like eating broccoli, but think of it as a workout and a relaxation that gives you energy. Have you ever wondered why you need to sit either with your legs tangled underneath you or slumped down low in a chair to be comfortable? Your body is trying to achieve the broad and stable base of support that it needs to be comfortable. You do not have to have big buttocks to have a broad base of support. You need to place your bones in the correct position, which will allow the muscles to relax.

4.1 Sitting Upright, Right B

Life Goal:

To be comfortable when you eat, work at your desk or when sitting in a backless chair.

Body Goal:

To improve hip range of motion, decrease stress on all aspects of the spine and strengthen the trunk muscles. Over time as the body adapts, less muscle work will be needed to sit correctly because the bones of the spine will be more ideally stacked. You will be a skilled sitter.

Props:

We did not grow up as many people in traditional societies did so we must use tools to help us return to our natural sitting position. A thick wedge or pillow that adapts as a wedge is a must in any chair that you sit on regularly. This will allow your body to relax and adapt. If you try to force the position without a wedge, it can create more problems so don't take this tool lightly.

Method:

Begin in a modified Butt Up the Wall position. Guide the tops of your thighs back with your fingertips.

With your knees bent and relaxed and your spine straight, think of the movement beginning from the seam of your pants (pelvic floor) and guide this area back and up.

A wedgeless jam? Move toward the edge of the seat and lean your upper body forward onto a table with a relaxed spine.

This will cause the rest of your spine to follow in line. Land on your pubic bone on the center of the wedge. For men, think not of sitting on your pants pockets but on the back of your thighs. For women, think of being in a hoop skirt and tossing the skirt over the back of the chair.

Once you're sitting, relax your inner thighs and stretch one leg out in front and the other under the chair so the thigh bones can rest in a slanted downward position on the chair. Your knees will be turned out and your ankle and foot only slightly turned out. The torso should feel like it is leaning forward slightly and the back is comfortable.

Your chest may now be elevated, so place one hand on the chest and the other at the base of your ribs to bring the chest down. Keep the spine straight as you sit upright. Keep the pelvis relaxed the entire time.

Note:

How do you know if you are sitting centered on the wedge? Close your eyes and lean slightly forward until you feel off center then lean back until you feel off center. Find the place between these two extremes; that is where you should be.

Places Without Wedges

There are times when you cannot create the perfect atmosphere for sitting. Here is the trick to get you out of a wedgeless jam: Go into the sitting position by bending deeply from the hips so you will land on the backs of the thighs and pubic bone. Move toward the edge of the seat and drop one knee down toward the floor and the other out in front. This will help the pelvis to naturally fall forward without straining the lower back. Then you can lean your upper body forward onto a table with a relaxed spine.

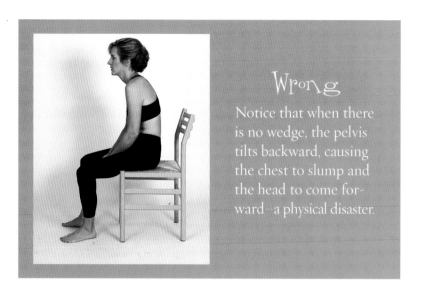

Wrong

Notice that when there is no wedge, the pelvis tilts backward, causing the chest to slump and the head to come forward—a physical disaster.

4.2 Recline Is Divine

Life Goal:

To enjoy sitting comfortably. Sit back at the movies, at the computer, in the car, on the plane, or the couch or at the bar. It can be divine if you learn how.

Props:

If you begin practicing this in a low-back chair or folding chair you will have a very easy time getting the feel.

Body Goal:

To relax tension in the shoulder, neck and lower back area while sitting. Sitting properly gives the spine information on what the correct position is by taking away holding patterns and creating a new important awareness.

Method:

Go into a sitting position as in Sitting Upright, Right (without a wedge). With your knees soft and spine straight, draw the seam of your pants back and up as you sit on your pubic bone towards the back of the chair.

Wrap a thin bath towel, sweatshirt or piece of fleece across your back and under the shoulder blades, holding the ends of the material in your hands. Let your chest sink and your upper back arch, making a "c" curve with your upper body, then lean back onto the chair against the towel.

Feel your shoulder blades supported by the towel. Take the ends of the towel and tuck them into your sides, parallel to your spine.

The towel should resemble a horseshoe, and should support you under your shoulder blades and parallel with your spine.

Lean back onto the material and bring your buttocks away from the back of the chair so the weight of your back on the chair falls against the upper spine, not your lumbar area. You know it is correct when it is comfortable.

4.2.1 Sitting With Your Legs Crossed

There are very few nevers in *Body Balance* but here is one. Do not cross your legs as in this picture (lady's pose). Instead, improve your hip range of motion and strength by crossing your legs with one ankle supported on your other knee. This can be a great alternative and help gain hip symmetry as well.

Method:

If you're sitting in a low-back chair forgo the towel or fleece, just hook your shoulder blades over the back of the chair.

Make sure your supporting leg is slanted slightly down so your hips can stay symmetrical. Begin crossing whichever leg seems most limited in movement to regain this hip's flexibility.

Do not hold this position so long that it creates discomfort. If you repeat the movement frequently, your hips will change for the better.

4.2.2 Getting Settled in Recline Is Divine

Support your head with your hands placed at the base of your skull. Let your shoulders drop away from your neck.

Relax your head into your hands, close your eyes and breathe in. As you exhale, let the weight of your head fall into your hands.

Go down your body and relax each area as you

exhale so that your chest, lower back, buttocks and thighs all relax toward the chair.

Your arms should be working to hold your head in place. Hold for 30 to 45 seconds.

Bring your arms down and notice how much more relaxed you are in the chair.

4.3 Spectator Stretch

Life Goal:

To make sitting a pleasant experience, whether you're at the ballpark, in a long meeting or waiting for a subway or plane.

Props:

You will need something to sit on that makes a wedge (pillow, jacket or magazines, depending on where you are).

Body Goal:

To regain natural hip flexibility and strengthen your back muscles by leaning forward and letting gravity work on the hip and spine.

Method:

Go into sitting as in Sitting Upright, Right. With your knees bent and relaxed and spine straight, take the seam of your pants back and up as you start to sit on your pubic bone on the support or wedge. Relax your pelvic floor, belly, groin and thigh muscles.

Place each foot at right angles to the knee. With one hand on the groove of the lower back (natural arch of the spine) and elbows facing back, begin to bend forward slowly keeping your spine straight while ensuring all the bending comes from your hips.

Allow your feet to move forward and out a little if needed to help relax your groin. Stop bending forward if you cannot maintain a groove.

Place your hands or forearms on your legs (depending on how far you can bend over) and rest. Keep your shoulders down, away from your neck. Relax into your pelvis, keeping your spine straight.

To come up, maintain a straight spine by lifting from the nape of your neck while relaxing the front of your body, especially your belly area. If necessary use your hands on your thighs to press yourself up so that you can maintain a relaxed pelvic area.

For comparison, notice when you are not bending correctly the spine curves forward at the chest and not at the hips Over time, this damages the spine and hips.

44 Get Up, Sit Up

Have you ever wondered where we got the idea of a sit-up? Its misuse and abuse creates more problems than answers. The sit-up itself, if done correctly, will bring the torso upright from the horizontal position. When doing this correctly, your trunk muscles (including your abdominals) will work as they were designed to.

Life Goal:

To learn how to sit on the floor comfortably with or without the support of your arms. Staying in this position for a period of time will work your back and abdominal muscles. Over time, it will not be difficult to sit like this.

Method:

Lie supine on floor and bend your knees. Place your feet flat on the floor about 15 inches apart. Bring your upper body up onto your forearms. Rotate the pelvis forward through your legs (let your stomach go) to create a soft arch in your low back.

Place your hands out to the sides with your palms down. Imagine someone lifting you up and forward by the back of your head. Using your hands, hinge the body up to the sit-up position.

Roll forward more onto your pubic bone. You will not be bending at your waist but you will have a small arch in your back. Now you're sitting upright on the floor.

Don't lock your elbows out, or bend at the waist.

Body Goal:

To strengthen the muscles along your spine and hips, as well as abdominals, while regaining natural hip movement.

4.5 Supported Sitting

Now that you're sitting up, this is a great workout for the upper back, arms and shoulder blade areas.

With your hands facing away from your body, allow the weight of your body to rest on your arms and hands. Unlock your elbows and press your hands into the floor to lift your torso. This will also take stress off your wrists and put much of the work on your upper back and arms.

Keep your chest from elevating by bringing your breastbone down to a more vertical position. Keep the back of your neck long by lifting from the nape of your neck or looking forward.

Keep the weight of your body back on your hands and keep your tummy rolled forward. This will create a small arch in your lower back area and should make your back feel good.

Get Up and Down to Get Around a

Life Goal:
To make getting up and down simpler.

Body Goal:
To gain functional strength and endurance.

Method:

Getting Down:

This is Butt Up the Wall with one hand out in front (much like a line-man in football).

Pivot on that hand and take your hips to the floor without bending at the waist. The key is to focus on the location of your buttocks the entire time.

Pelvic Position:

For all the sitting pos-tures the pelvis will begin as it does in the standing posture, then move into the position of Butt Up the Wall, and end in the posi-tion that is the same as the standing posture.

Hinge to the upright position as in coming out of Butt Up the Wall.

You will be in a pseudo Butt Up the Wall position.

Turn your torso as a whole toward the side of the leg that is down. Focus on leading with your belly button. With the weight on your hand on the same side, and without thinking too much, just take your buttocks to the sky.

Getting Up. Start in the Get Up, Sit Up position. Place one leg under the other (as in picture). Keep your torso open.

Getting Up:

Belly dancers are a natural at the rib tilt. You now can use this subtle but powerful move to lift the weight off your back and get toned around the middle at the same time.

5.1 Rib Tilt in Sitting B

Life Goal:

To realign the mid-spine with gravity. You will notice a change in the position of your rib cage. Your chest will be open and your neck will be back over your shoulders.

Note:

Your abdominal area will feel lengthened and tighter. Your waistline will feel lengthened and rotation around the waist will be easy. Rib tilting can be done sitting, standing and, eventually, even walking. You are learning the Rib Tilt passively as you sit in Recline is Divine because gravity never stops working. It's constantly making adjustments without your help.

Body Goal:

To help you decrease pressure in your lower back by allowing your trunk muscles, including your abdominals, to function more efficiently.

Method:

Sitting Upright Right put your arms out to the side, palms up.

Now place one hand at the base of your ribs. Then draw the ribs down, tilting your body slightly forward at the waist.

Bring the back of your neck upright and keep the pelvis in place.

Note:

It only takes the slightest tip to get a change. Think of tipping your ribs.

Pelvic Position:

The pelvis is in the same position as in sitting and standing.

5.2 Marionette i

Life Goal:

To enable you to carry luggage, briefcases, children and shopping bags without your neck and back letting you down.

Props:
Wall

Body Goal:

To increase core stability and awareness of your mid-spine, relieving low-back and neck tension.

Note:

We often feel a need to stretch forward in the upper body like cats. Learning the *Body Balance* method, your spine is changing to a more upright position so sometimes it feels like it needs to go the other way and flex forward, which is natural and important. Doing this can help you in other moves such as Recline Is Divine. It is an exaggerated Rib Tilt and over time you lose the tightness in the lower back and shoulder blades as your body just stacks up better. It's a great body break if you've been sitting at your desk for a while.

Method:

Pelvic Position:

The pelvis is tilted back instead of forward because you are hanging on the wall. This is a relaxed position, and the spine loses the lower arch naturally.

Begin close to a wall. Come up on your tiptoes and arch your upper spine forward like a cat, allowing your shoulders to rise as your pelvis relaxes (you will be bending at the waist).

Place your mid-back (thoracic) on the wall as if it were hanging like a marionette, and then relax your shoulders, neck and head. Once there, the more you relax the easier it will be to hang. Hang from that spot where your back is on the wall. Stay on your toes and relax the rest of your body, hanging. Your lower back will be hanging straight without a groove. You may not feel in the beginning as if you are pressing your back into the wall while hanging. It will be very subtle. In time you will feel the isolation of hanging from your mid-thoracic.

Notice that this Portuguese fisherman is not a young man and that he has good muscle development for his age. His strong upright posture tells us he is active daily. You may move less than he does in a day, but you can regain your natural alignment with the Body Balance principles.

Shoulders

The skeletal structure is responsible for creating the look of an open chest and broad shoulders. It can't be built by weight lifting; it's how the structure naturally hangs that catches your eye. So throw those weights in the bag, and learn how to carry them, which will give you the workout that will really fill the shirt.

Method:

Sit in the Upright, Right position on a wedge.

Pelvic Position:

The pelvis is in the same position as it is in sitting and standing.

Lift your arms out from your sides with your palms up, gently without forcing the position. Move your arms down and out from the armpit, keeping the top part of the shoulder relaxed.

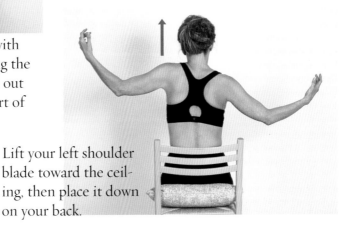

Lift your left shoulder blade toward the ceiling, then place it down on your back.

6. Shoulder Opener   i

Life Goal:

To carry items and become stronger. Carrying groceries, luggage, children, a purse or briefcase can be a great way to get the upper body strong and upright.

Body Goal:

To help the upper spine gain its natural alignment, allowing the postural muscles to regain strength.

Props:

A wedge.

Lift your right shoulder up towards the ceiling, then place that shoulder blade down on your back.

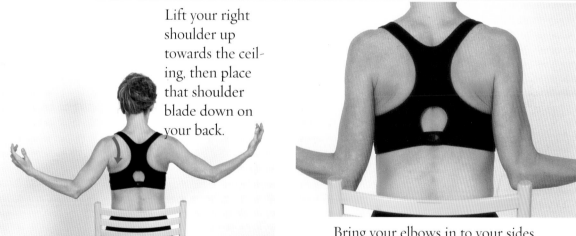

Bring your elbows in to your sides.

Do a Rib Tilt by placing your hands on your chest and bring your ribs down.

The Neck

This man from Morocco is illustrating how far the head and neck should be able to bend forward naturally without his shoulders slumping forward. The vertical lines on his robe illustrate his upright posture.

7.1 Turtle Neck

Life Goal:

To decrease tension and pain in neck and shoulders.

Pelvic Position:

The pelvis is in the same position as in sitting and standing.

Body Goal:

To get the neck back in line with the rest of the spine.

Never

do Turtle Neck without first setting up the shoulder blades and keeping them in place. There should be no pain or blocking of movement in the neck. If there is, stop the movement. It should feel like a stretch in the back of your neck or between your shoulder blades. It should feel good.

Repeat

Do it with your eyes closed to get a better feeling for the movement. Go very slowly so your body can communicate the different feelings, and the movement will have a more lasting impact.

Sit upright on a wedge and begin with a Shoulder Opener. Keeping your shoulder blades on your back, bring your chest down in front.

Bring your arms into the sides of your body. Without allowing your shoulder blades to move off your back, bring your face forward, out and slightly down, leading with your nose. Do not lift your face up or down but go directly out at an angle towards the ground.

THEN bring the back of your neck back over your shoulders by thinking about elongating the BACK of the neck. Your shoulder blades remain active and down as counter-resistance.

Repetitions:

Do no more than 2 or 3 repetitions at a time; focus on breathing and keeping the pelvic area relaxed while doing the moves.

7.2 Feel Your Neck B

A lot of the times in daily movement the neck can get in line through passive placements. This means leaning back and resting your head in the right place in the car or plane or on the couch at home.

Our culture encourages the forward-head posture because so much of our life demands that our upper body be in static positions for long periods of time. This can cause carpal tunnel syndrome and tension headaches.

The key to having a relaxed, strong neck is to touch it. The neck is the first part of the spine to get out of line. The more hands-on we are with this area of our body, the quicker the changes occur to give us a more alert, relaxed and comfortable state.

Your neck needs constant feedback to correct problems because it is so position-sensitive. It needs minimal resistance and a lot of correct repeated moves. It learns best by following your trunk around in the correct position.

Every move in Body Balance can help your neck, but always check it with your hands to make sure it is not ahead of the game. A touch on your neck says more to your body than a thousand words.

Place one hand on your neck. (Note: if you have long hair you can hold your hair up with your free hand.)

Close your eyes and relax your shoulders, chest and belly.

Bring your head slightly forward, leading with your nose. Do not lift your face up but go directly out at an angle towards the ground. Feel how ropy and thin the muscles become under your hand. This is not the natural way your neck should feel.

Bring your neck back into your hand while you allow your chin to drop. Now grab the base of your hair at your nape and gently lift.

Do not think about working your face and jaw muscles. Just keep your mind on your neck while you broaden it into your hand.

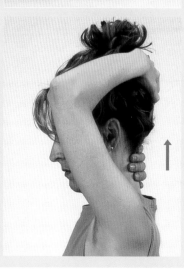

This little awareness trick can be done any time to regain natural alignment.

Floor | Shoulders

Standing | Lying

Ribs | Neck

Feet | Bending

Tricks | Balancing

Sitting

Standing on one leg does not seem to be a big deal; it is how most of us are comfortable standing. It can be either hard on the joints and ligaments or a great stimulus to building stronger bones and a better stride. Gravity never takes a vacation. Even when you think you are not working, it's working on you. So, repeatedly standing incorrectly wears out your joints and ligaments.

8.1 Bend and Balance

Life Goal:

To stand and bend while balancing on one leg. To walk with more power and less effort.

Body Goal:

To gain functional strength in the back, buttocks and hamstring muscles. To gain better balance in all moves.

Props:

Have a tall ledge or chair to lean on for support.

This is the one move that brings all of the things you have learned from the other moves together. Don't forget that all movements build upon themselves, and when these moves are integrated into your daily life, all movements are easier.

Make sure your first toe is planted firmly with your knees soft (bent) and turned out towards the outer toes.

Place one hand in the groove of your back and begin to go into Butt Up the Wall on an imaginary wall.

Now shift your weight to your left heel and buttocks while keeping your hips level. Keep your left leg rotated out so the weight is in the back, not side, of your right hip. Slide your right foot next to your left.

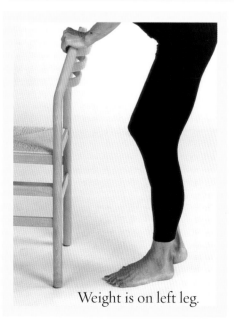

Weight is on left leg.

Adjust your shoulder blades onto your back. You can slide your right leg back but keep the toes on the floor.

Imagine placing a tea cup on the base of your spine at the sacrum. Your tea cup may be at a slant down toward the floor in the beginning but should not be tipping to the sides.

Hold this position for a relaxing breath cycle.

Come up by pressing your left heel to the floor, feeling the connection between the heel and the buttock on your left side. Think of lifting the back of your neck up as you come upright with your spine straight. Keep the weight on the back of your right leg and hip as you come up with weight on the left only to maintain balance.

Repeat on your right leg.

Repetitions:

Do this slowly only 1 to 3 times while focusing on the technique.

The feel

The backs of your legs (and maybe your calf and buttock muscles) may feel stretched, but you should not feel sharp pain. It may feel like a lot of work is being done in the leg that is holding most of the weight.

Lighten this sensation by putting more weight on your arms. If there is weight on the front of your knee, straighten your leg a little and shift your weight back into your buttocks by visualizing Butt Up the Wall until the weight or pressure is off of your knee.

This is how it should feel naturally standing on your right foot. Take a mental picture of how this feels and how the left leg is not involved in weight bearing as much.

Note:

Want to work more on balance? Let go of the ledge as you come up, focusing on pressure in the heel of the leg you are balancing on and its connection to the buttock. Continue to lift the spine from the neck as well.

Pelvic Position:

The pelvis is in the same position as in standing in the beginning. Then it moves into the beginning of a Butt Up the Wall position.

Did you know that walking is a form of balancing on one leg and the better you are at walking the better your balance will be?

Apply to life

Since this is a rather involved move, it can be modified and applied to daily life easily. When you notice yourself standing in your usual posture, just weight shift to the new way. Lean slightly forward from your hip. Keep your knee relaxed and shift your weight to your buttocks with your other foot relaxed but under your body. Bring your body upright from the nape of the neck as you relax your shoulders and belly.

8.2 Walking

Begin as if you were going into the Standing Pose by leaning slightly forward at your hips with your hands behind you. Your buttocks will be behind you instead of tucked.

Swing, don't place, one leg forward, letting the heel land naturally.

Now, give a big push off with the leg you are standing on, feeling as though the push off is throwing you forward.

Walking

More than any other move, walking will continue to show noticeable improvement over time as you get the physical language of the other Universal Movements. Walking is a reflection of our posture and how our body understands the use of all the joints.

Helpful tips

Look at top runners and observe how they accelerate from the way their legs extend dramatically behind them without changing the position of their pelvis.

Basic principles

Start by studying your natural walk. Get to know how your pelvis is positioned and whether you push or pull to get across the room. Using a mirror can help.

Helpful tips:

Over time, you will not have to lean so far forward, but in the beginning, to get the feel, you must exaggerate the move.

Continue with your focus on the leg that is responsible for pushing you forward. Think of using the whole foot to accelerate forward as you peel it off the ground behind you to exaggerate the feel of it propelling you forward. Do not land on your toes, though it will be difficult in the beginning not to because of the momentum of the push off.

Think of the power of the push off coming from the buttocks, hamstring and calf muscles.

Compare these two series of photos. Notice below when your buttocks are tucked how it adversely affects the line of the body. The posture above is more energy efficient, not to mention aesthetically more pleasing.

Your walk is a reflection of your posture.

8.2.1 Partner Power for Power Walking B

Pelvic Position:

The pelvis is positioned first in the standing posture then in a modified Butt Up the Wall position.

Body Balance moves

are about focusing on technique and repetition, not forcing the issue on the tissue.

Partner Power

Using resistance will help you sense a change in the way you walk. Let a partner be the object that you are pushing.

Set up your torso by going into Standing Pose by leaning slightly forward at your hips. Use your partner's hands and body as a resistance.

Your partner walks backwards resisting you. This helps you regain the feeling of pushing off with the leg that is behind you. The other leg should feel more relaxed and not involved in the forward motion until it has crossed the midline of your body.

When you begin to get the feel of walking with more ease, you will notice that your balance will improve because of the position of your pelvis. Combine Feel Your Neck and Rib Tilt with the walk.

If you are in acute pain or just cannot get rid of an ache, the secret cure may be in doing nothing, but doing it well.

Lying

Horizontal Healing is about comfort moves or passive exercises. They are my favorites. Sometimes we try too hard to get changes. Here, you just set the body up and let it all happen with gravity doing the workout on you.

It is very important to realize that being comfort-able in these positions is the key to their working. The setup is everything. Stay as long as you like. The body settles like soil, and after a short time, your body will feel even better if you move it again to a more comfortable level.

9.1 Frog Pose B

Props:

A thick blanket, pillow or fleece folded lengthwise. It should end up 6 to 8 inches wide and the length of your torso.

Body Goal:

To gain awareness of how the front of the body can be more open. Gravity does the work by helping relax your shoulder and hip joints, where a lot of holding patterns occur. Since frog is a passive exercise, focus on breathing deep toward your lower back area, allowing more dynamic changes to occur.

Do not do it

if this pose is not comfortable for you. Make sure you follow the guidelines carefully because lying prone without the correct support can put improper stress on joints and lead to painful problems.

Pelvic Position:

The pelvis is positioned so there is a natural arch in the spine as in all the positions throughout this chapter. Avoid having too much arch in the lower back spine by building up the support under the body.

Place the material or pillows on the floor. Lie prone (on your tummy) on the material.

Bring your hands to your ribcage and draw the front of your ribs down (the same way you would bring a shirt down in front of your body).

Rotate your thighs so your inner thighs are on the pillow and your lower legs and feet are turned out.

Allow your head to be slightly tilted down towards the floor and supported with the material you are lying on.

You may need to build the material higher if your hips or low back do not feel comfortable.

Each time you do this, the height of the material may change due to the changes in your body's comfort.

9.2 Supine Feels Fine B

Props:
2 pillows.

Body Goal:
To regain hip range of motion, decrease the stress on your back while lying and bring the neck back to its natural alignment. It is also a sleeping position.

Lie supine with your pillow slightly under your shoulder blades, creating a wedge for your neck. Make sure it is comfortable but not too high. You can relax one leg out on a pillow with the foot close to the calf of the other leg. The support should be high enough that the leg relaxes. (The body positioning principles are the same as those used in the Floor Moves, Rock and Roll.) Place the leg that feels tighter on the pillow to regain natural movement of that hip. Now breathe, relax and let your body transform as you sleep.

Neck Rolls

Pillow Under Knees

DO NOT USE PILLOWS UNDER YOUR KNEES to take the pressure off the back. This puts the hamstrings in a shortened position and your body in a pelvic tilt. It may feel great—temporarily—but it is creating adaptations that over time will throw your body more out of balance the more it is done.

DO NOT USE NECK ROLLS. They are bad for the neck. With a pillow positioned as a wedge and a little support in the neck, you will be able to gradually have less pillow and a more upright shoulder-neck posture because these support systems encourage the body to go back to its natural alignment. They do not support your body in the dysfunctional pattern you are already in.

9.2 Side Lying or Sleeping Position B

Life Goal:

To increase the benefits of sleeping restfully. If you have trouble getting a good night's sleep, positioning your body correctly could help you get more ZZZZs. Small adjustments can make big changes, so take each little movement seriously.

Pelvic Position:

The pelvis is in the same position as sitting. The buttocks are back, not tucked.

The example will illustrate sleeping on the right side, but sleep on the side that seems most natural or comfortable.

Lie on your right side with a pillow under your head. Bring your right shoulder or arm slightly forward until you feel you are not lying on the shoulder bone, but a little on the back of the shoulder.

Place the fingers of your left hand under the right hip bone in front and push your hip back and away from your waistline.

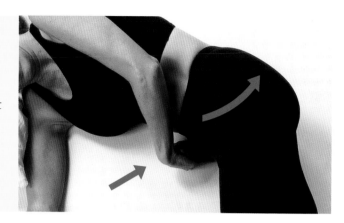

Your buttocks are now behind you rather than tucked, and your spine should feel lengthened.

Place a pillow between your legs with your legs slightly bent to a comfortable position. (If you support your foot as well this can bring added comfort if you have hip or knee pain.)

Tuck your ribs down in front as your upper back flexes forward. Adjust your neck by tilting your face down.

Rest your left arm on your side, on a pillow in front of you, or wherever it is comfortable.

When your big toe pushes into your second and third toes, the toes look crammed. The knuckle of the first toe is enlarged and irritated, which often creates a bunion. Much of this is due to unnecessary wear and tear on the joint because of the way your weight is distributed through your body and onto your feet. Surgery is popular for this problem but, in many cases, it is probably unnecessary and unwise because of the first toe joint's importance for balance and movement.

When you learn to redistribute your weight throughout your foot correctly and use your foot in a more functional way, you can reawaken the sensory receptors so the foot can respond correctly for balance, stability and mobility.

10 Get a grip B

Repeat

Do this exercise until it is not difficult and you can stand naturally with less turnout of your foot.

Body Goal:

To regain functional strength of your feet, and to create better alignment with your knees and hips in movement. To help correct over-pronation problems in feet.

Props:

A chair.

Method:

Sit in Upright, Right with your feet resting on the floor 15 inches apart.

Turn your toes up, press down with the knuckle of the first toe and then slowly relax and lower all the other toes. Keep the pressure on the knuckle of your first toe without crunching your toes.

Now slowly allow your legs to press apart without losing the pressure and position of the knuckle of each first toe. Keep a balance between the two, and do not allow your legs to open more if your toes lose their position.

Maintain this position for a few minutes if possible, then rest.

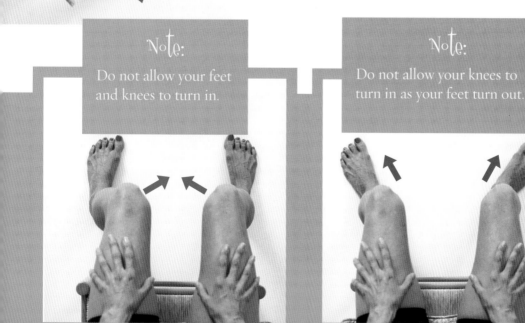

Note:
Do not allow your feet and knees to turn in.

Note:
Do not allow your knees to turn in as your feet turn out.

Floor	Shoulders
Standing	Lying
Ribs	Neck
Feet	Balancing
Bending	Tricks
Sitting	

These are moves that are important but don't seem to fall into any one category; but they help you learn the art of moving just as well.

11.1 Pulling Weeds i

Props:

Support for the knee and a chair. Mirror, if possible.

Body Goal:

To help you gain functional range of motion of the hips, which is a direct influence on gaining strength in your back, buttocks and legs.

Life Goal:

To help you kneel and work with your hands close to the floor. This is a good position for gardening or any activity close to the ground. This can transform the typical groin stretch into a daily activity and over time the groin muscles will be functioning without that tight feeling. You will no longer have to fear pulling it if you move your hips and legs as they were designed.

Begin movement by kneeling on your right knee with a soft support for the knee. Your left leg is bent with the foot on the ground and knee out slightly from the center of your body. Position a chair in front of body for support.

Place your hands on your hips and allow your left groin to relax as your hip drops to become level. With your spine straight, slowly begin to rotate your pelvis forward maintaining a groove in your lower back area (Butt Up the Wall).

Don't bend at the waist as in these pictures.

Rotate a very short distance. You will begin to feel a stretch throughout the right groin and hip. Stop rotating. Rest your upper body forward on the chair, keeping your shoulder blades down on the back, and relax into your groin. Do not force; just allow the body to rest in this position.

As you maintain the stretch while relaxing, slowly begin to guide your body back and forth. Continue to relax. Repeat this back and forward movement, feeling the tension in your hamstrings or groin area. Continue until the area feels more relaxed or tired.

This tension will decrease over time through adaptation, and the movement will become more comfortable if you integrate it into your daily life. Do two to three times, breathing five to six breath cycles per leg. Be sure to integrate this into daily life.

Grooves

What's a Groovy Groove:

A groove may be deep or shallow depending upon how much you use your back. The goal is an EVEN groove up the spine, not deep in one area and shallow in another.

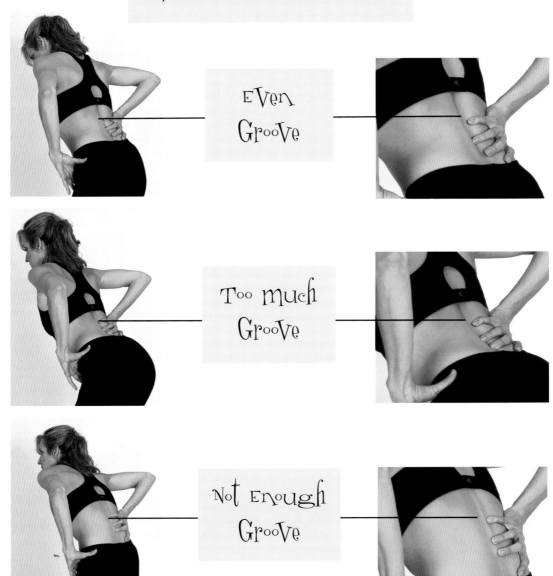

EVEN Groove

Too much Groove

Not Enough Groove

11.2 Torso Turning

Stand.
Place a hand on your lower abdominal area and relax into your hand.

Leading with the hand on your abdominals, turn your torso. Allow the rest of your body to follow. The more you think of leading with this area of your body when turning, the more the rest of your spine and legs will respond naturally.

You will be moving your feet more instead of twisting your trunk with your feet planted. It is much like dancing.

Do not twist

your neck or your trunk with your feet planted when turning.

Pelvic Position:

The pelvis is positioned the same as in standing and walking.

Simplicity

This is one of the simplest but most effective principles to learn.

Lifestyle Workout Examples

Following are sample ideas of how to put the Body Balance movements together in a routine. However, I encourage you to create your own routine for your lifestyle. For instance, if you know you are going to walk, shop or sit at a desk, build these workouts around your activity.

Starter Kit

Wake up and do a couple of Rock and Rolls in bed, including Rib Tilt, just to get your joints happy.

While you're splashing your face or brushing your teeth, bend with Butt Up the Wall.

Pick up your newspaper with Butt Up the Wall.

Get your orange juice, or coffee, and sit on a wedge—Sitting Upright Right— while you eat or read the newspaper.

If you're wearing trousers, put them on with another Butt Up the Wall.

Keep up the good work while you put on socks or shoes. Sit on the edge of your bed, with a pillow as a wedge, keep your spine straight and cross one leg so your ankle is on your thigh. Put on your shoe. Make sure your spine is as straight as possible.

To tie or buckle your shoe, go into Spectator Stretch. You may feel a stretch in your hip at first, but you'll gain your natural movement over time.

Now it's time to take out the trash. Pick up the trash can doing Butt Up the Wall, Shoulder Openers and Rib Tilt; walk with the push off, bend over with Butt Up the Wall again to put the trash can down, keeping those Shoulder Openers and Rib Tilts active.

Pick up your briefcase carrying it with a Shoulder Opener and don't forget that neck position.

Put the briefcase in the car (Butt Up the Wall), get in (Butt Up the Wall), sit in Recline Is Divine (use your towel) and rest that neck as you drive to work.

You can skip your stretch class tonight. Now your swimming, tennis game, racquetball game or other sport will be your added strength training, so you can wave as you pass by the weight room.

The Office Worker or New York Walker

While you are sitting in Sitting Upright Right, you get a telephone call. It's going to be long, so put the call on speaker phone and get into Spectator Stretch with your papers on the floor in front of you.

The call is even longer than you thought and tension is rising. You now pace the room, walking with the *Body Balance* principles. You've walked off that tension, so sit back into Recline Is Divine. This rests your mind and your thinking is better. The deal is done.

It's lunchtime. Power Walk down the hall, get into the elevator and do Butt Up the Wall, or take the stairs using the principles of Butt Up the Wall to sneak in a little workout.

It's 10 blocks to the restaurant. It used to take 20 minutes but now it takes 10.

After lunch, sit in Recline Is Divine. Take in that moment of rest.

At the end of the day you pick up your briefcase and know your weight workout has started.

Walkers

Depending on your time and schedule before your walk, start with floor moves or Butt Up the Wall. This gets your awareness honed for that walk. Take the walking principles with you including the Rib Tilt and Shoulder Openers.

When you've finished, come in and do Butt Up the Wall and Pulling the Weeds. Get a glass of water, and go into Up and Down to Get Around and get on the floor. Sit in Get Up, Sit Up so you can get your upper body workout while you watch your favorite news broadcast.

Roll over and rest in Side Lying because you deserve it and your body needs to know how to relax.

The Suburban Mom

You're a manual laborer and taxi driver, and you work long hours.

8:30 a.m.   Pack the kids in the car; get behind the wheel in Recline Is Divine and you're ready for car pool.

10:00 a.m.   The kids are at school and there are still six more errands to run. You pop in and out of that Suburban with the knowledge of Butt Up the Wall. Carrying things is a no-brainer as you do Shoulder Openers, Rib Tilts, Standing in Line.

12 noon   Lunch with the girls as you sit on your sweatshirt (wedge) in Sit Upright Right.

2:00 p.m.   Time to get the groceries before you pick up the kids for soccer practice. You carry the groceries because you want that good workout you get with Shoulder Openers, Rib Tilts, Standing in Line. You load the sacks in the Suburban with Butt Up the Wall.

3:15 p.m.   You jump into the Suburban and sit in Recline Is Divine. It's car pool time and you take the kids to soccer practice while you go to your tennis lesson.

3:40 p.m.   Your tennis pro asks "Why are looking so fresh and rested?"

"I move all day with
The McCall Body Balance Method!!"

CHAPTER V
The REAL PEOPLE

CHAPTER V

THE REAL PEOPLE

Here, a simple couch pillow helps create a smaller "couch." The support of the pillow is not in the low back area but at the upper spine, directly below her shoulder blades.

To truly treat injuries through movement, the method used must apply equally to all people.

You can get your body working more comfortably and efficiently with *Body Balance*, whether you're a child, a teenager, a professional athlete, or a person with special physical challenges. This chapter gives case histories from people in all walks of life.

Why We Must Start 'Em Young

You may wake up one day and say, "Wow, my body hurts!" But it's generally not something that has just happened that causes you this pain. It's like the national debt, crime rates or pollution. You have to ignore the evidence and misuse the resources for a long time for a problem "suddenly" to appear.

Today's technology-driven world offers so much, but also has created a new set of physical problems. The wonders of the Web, and computers in general, have taken us into a world that focuses on being entertained instead of entertaining ourselves with our own creative minds and movements. Since parents fear crime they encourage children not to play in their own neighborhood streets, which leads often to more sedentary indoor activities. Lost are the running, jumping, falling and physical contacts that teach children how to move naturally and safely. Many children not outside playing games will probably be in organized activities and sports. We've actually become a society that needs "organized playtime." One consequence of this

Rachel unknowingly slumps as she reads. Over time this habitual posture can make changes along her spine and hips, setting her up for injuries.

structured recreation is that children don't learn important basic movements that need to be a part of physical learning at a very early stage before structured contact sports begin.

For instance, young athletes (male or female) should be taught how to fall early in life. This is one of those important movements that will keep them playing sports longer and healthier. To understand this, look at wide receivers in football. They jump, get hit and fall. Then they get back up, time and time again, because they know how to fall.

Many of us didn't realize when we were young that during those after-school games of tag or pickup football, we were really learning a skill that kids don't get today.

On the flip side, these more structured sports have given girls greater opportunities to get involved in team sports. These sports—such as soccer, basketball, volley-

A sports medicine think-tank confirms this:
"... in the past, youngsters were encouraged to do outdoor activities like playing tag, running, jumping from a jungle gym .... No winning, just fun. Playing outside was fun. Now individuals often are playing inside, using computers and watching television. The movement patterns and ability to fall should be taught early in life."

ball, swimming and hockey—are ways of developing needed movement and healthy neuromuscular systems. However, these young people can slip into the same "body" trap as their parents. If our movement patterns are drastically different when we're on and off the field, they set us up for injuries.

Human beings are great imitators. That's how we learn. Look at the picture on page 65 of the child carrying sticks on her head. You can imagine her learning how to best do this by watching the adults around her.

In our culture, children do the same thing. How can we make a simple change right now for a young person doing a typical activity? On the opposite page there are two pictures of Rachel reading. By simply changing the way she sits, she makes positive changes to her body, instead of negative ones. When she sits using the *Body Balance* principles, her body will adapt to the correct way. Our bodies are always adapting, correctly or incorrectly. She maintains her natural alignment and will not have to relearn how to sit as she gets older. At this age, children still haven't lost that critical range of motion in their hips and spines so their bodies can learn and adapt comfortably in the way they were designed.

Kelly's Question

When I gave a seminar for athletic trainers-in-training at Edinboro, Pennsylvania, Kelly conducted her own instant experiment with *Body Balance* work.

During the first evening I showed Kelly how to lean back in a chair correctly and comfortably and how to stand. The next day she walked into the athletic training room with a new intensity in her eyes. This was a different Kelly. Both her body and her face had changed. She seemed taller and more alive than she had just the night before.

"I tested your work today in class," she told me. "I sat like you showed me for 30 minutes, then for the last part of class I 'rewarded' myself by sitting as I wanted, my old way.

"When I sat like you said, it was the first time I heard everything the professor said. The ideas went in my brain and on the paper. Usually I am looking at my watch and thinking about what I will be doing later. The last 10 minutes, when I sat in my usual position, I almost fell asleep—as usual."

Kelly had changed mentally and physically quickly and without much effort. It was the combination of both her mind's and body's responses to *Body Balance* that brought such dramatic results. This 19-year-old impressed me. Kelly was a very observant trainee as an athletic trainer who was also studying to be an elementary teacher.

The previous day she'd posed this vital question: "Why are these changes occurring so quickly? I have watched you for two days; what are you doing to make such quick changes in these athletes?"

I explained that as a society we had lost the physical intelligence of how to move. Young people, because of their environment, advertisements and parents with the same problems, were losing their natural, normal movement. This translates over time into many musculoskeletal problems that develop in life, from loss of mobility to chronic back pain.

Her response was insightful. "Everyone should know this stuff. Why aren't we teaching this to kids in elementary school where it all begins?"

This book, I hope, will help Kelly, and other like-minded people, to take these movement principles, learn them and give them back to younger people.

Anna's Story

After spending an hour with Anna you have to remind yourself that you are talking to someone who is only 20. Her maturity, spark for life and deep determination are impressive for her age.

Give Anna responsibility and you know you'll get a job done well. She goes after most things with a sensitive, conscientious style. This attitude shows up in her sports. But a few years ago, Anna did not know where to turn to find the answer to her knee and back problems.

In high school, she played sports every spare minute she had. Field hockey, soccer, softball, you name it, she played it. Then, in her junior year in high school, a knee pain developed that could not be diagnosed. She finally found a therapist who helped her by doing spinal adjustments and, to her surprise, the knee pain went away. She ran in 5-kilometer races during her first two years in college and everything seemed fine until the adjustments had to be more frequent. Progressive back pain followed, with the adjustments failing to bring relief.

Three years later, when I met Anna, her knee and back pain had kept her from working out for a full semester and now weight gain was an added concern. At this point, simply bending and picking up a backpack was a nightmare. No form of exercise seemed possible because her back would "go out." But she could not imagine life without playing sports.

Most of Anna's joints had too much mobility or "joint play" and she needed to keep them under her control. After a dedicated few months of *Body Balance* class, she started working out on a ski machine preparing for a trip abroad. That summer she went to school abroad and roughed it through Europe—hauling the backpack that used to give her nightmares. Two years later she toured Europe again and did even better.

Today, she loves spinning class (on stationary bicycles) and is overjoyed to be sweating again like a true athlete. The pounds are falling off, and she has become an avid tennis player.

How did she get in such a mess at a young age? She did not know how to move her body. She allowed someone to "adjust" her back but never learned how to keep it there.

Anna is a perfect example of how a lack of intelligent movement sets up young people in our society for injuries that could keep them sidelined for life.

Cutting On the Curve: Scoliosis

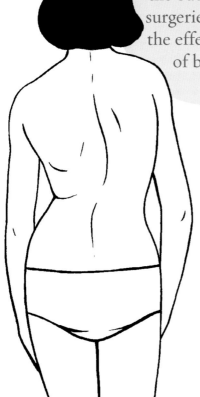

Scoliosis itself does not progress in a consistent way, nor is there a pattern to the outcomes of surgeries or even the effectiveness of braces.

Young people with scoliosis need to regain their physical intelligence to cope with their curved spines. This side-to-side spinal curve looks much like an S or a C.[2] It is really a three-dimensional curve, and rotation of some of the vertebrae is common. The shoulders and hips of many scoliosis sufferers actually look uneven. In advanced cases, the thoracic area will be affected, causing the ribs to shift as the spine "curls" toward the concave side.[3]

The causes of 80 percent of the cases are unknown.[4] Scoliosis usually appears in mid- or late childhood, prior to puberty, and is seen more in girls than in boys.[5] In mild cases, the curve can range from 5 to 15 degrees while severe cases show curves of more than 45 degrees.

The most common forms of treatment are bracing and surgery. Bracing is recommended for curves that are 20 to 40 degrees out of alignment, but only for adolescents because their bones are still growing. A brace is worn most of the day for a number of years, depending on the degree of the curve and bone maturity.[6] This is an emotional and psychological burden for young people.

If the curve progresses beyond 40 or 50 degrees, surgery is the choice of treatment.[7] This is done sometimes to help relieve pain and sometimes as a preventive measure because of fears that the progression of the curve will create cardiac or respiratory problems.[8] There are five commonly used surg-

The immense mechanical power of the body is demonstrated by the fact that often the rods stress and break. Sadly, this can lead to more extensive surgery.

eries to correct scoliosis. One that is frequently used is fusing segments of the spine together (usually 10 segments) then straightening the spine with stainless steel rods, called Harrington rods.[9] It is very extensive and expensive surgery, and there are no guarantees that it will stop the pain.

The immense mechanical power of the body is demonstrated by the fact that often the rods stress and break. Sadly, this can lead to more extensive surgery.

Until now, therapeutic exercise has not been effective in slowing the progression of the curve or in preventing pain. Overall, there is little help available to scoliosis patients aside from bracing or surgery. Scoliosis itself does not progress in a consistent way, nor is there a pattern to the outcomes of surgeries or even the effectiveness of braces.

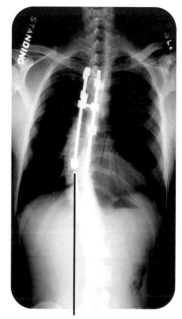

Broken Rod

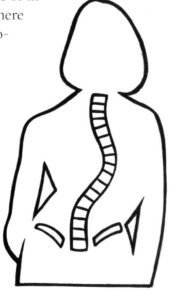

. . . the Milwaukee brace (one of the most commonly used) keeps the pelvis in an unnatural position for up to 23 hours a day.

Using Untapped Resources

The *Body Balance* movements work the neuromuscular system unlike other formal exercises. A brace only puts up a wall to stop the spine whereas *Body Balance* movements work like a natural brace.

For instance, the Milwaukee brace (one of the most commonly used) keeps the pelvis in an unnatural position for up to 23 hours a day. It also limits the power of the muscles to aid in counteracting the curve.

On the other hand, *Body Balance* moves powerfully stimulate the neuromuscular system through daily activities. Combining these moves with a brace that is less restrictive in the pelvic region may give a more effective and less traumatic outcome.

Our bodies were designed to fix themselves whenever possible. However, in this immobile society, we have lost the knowledge of how normal everyday movement can correct musculoskeletal problems. Reclaiming correct movement allows the body to help itself.

Body Balance moves build motor engrams (memories of movements) needed to help the body act on its own. We must continue to move our pelvis and spine because it's the only way the body learns how to move properly.

Bones and muscles depend on external mechanical forces to stimulate change. Consistent bombardment of the musculoskeletal system with the correct stimulus can create positive change.

If the body has the ability to break steel rods, it has the ability to help develop a healthier spine. There is a place for surgery and braces. But, as I have seen with my scoliosis patients, profound changes can be made with *Body Balance* movements. These moves not only help patients avoid surgery, but also help them recover more quickly from surgery, when that is the choice.

Body Balance has helped people whose surgery was unsuccessful as well as adults with severe curves who did not have surgery. Adolescents, with their young neuromuscular systems, can make more aggressive changes because of their tissue pliability.

X-Rated Side Effect

A harmful side effect of monitoring scoliosis, particularly for young women, is the use of X-rays and their link with breast cancer. X-rays are used at regular intervals during important years of development to monitor the curve's progression.[10] One study in 1979 postulated that a girl would receive an average of 22 X-rays during her growth years, which would increase her risk of breast cancer by 110 percent.

Nowadays, more care is being taken to use breast and pelvic shields and there are non-radiation methods, such as moiré topography or Scoliometers, to monitor scoliosis.[11]

In Their Own Words

When surgery is the choice of treatment, patients face the problem common with any fusion of the spine: the segments above and below take on an unnatural load. Through fusion, the spine has been turned into a rod, not the spring it was designed to be. Incorrect forces on the adjacent segments cause wear and tear of these segments and may lead to joint breakdown, such as degenerative joint disease.

Before you choose your surgeon, see how willing he or she is to look at alternative treatments.

Body Balance A Natural Brace

A brace only puts up a wall to stop the spine whereas *Body Balance* movements work like a natural brace.

Listen to these strong advocates of surgery and judge their approaches.

1. *Dr. James Ogilvie, an orthopedic surgeon, in the book* Stopping Scoliosis *which is recommended by the Scoliosis Foundation, states:* "Voluntary exercise alone will never improve or eliminate a curve, because the brain cannot command specific paraspinal muscles [muscles next to the spine] to contract and relax. It is those muscles and surrounding tissues, when strong and healthy, that keep a spine in a straight position. When they're weak on one side or the other, the spine will bend. And even if the brain could make the right connections, who would have the discipline or stamina necessary to carry out such an exercise program?"[12]

2. *The author of* Stopping Scoliosis, *Nancy Schommer, also refers to Dr. Hugo Keim's book,* The Adolescent Spine *saying:* "In the second edition of this widely used reference book, he makes an extremely persuasive case against exercise." *Schommer then quotes from Dr. Keim's book:* "They [exercises] maintain and enhance body tone and are of value to the patient and family because they make the parents feel that they are doing something, which assuages their guilt feelings somewhat.... To sum up the indications for an exercise program, you can prescribe it if you wish, as long as you understand that exercises only treat the psyches of the parents and help the muscle coordination of certain poorly muscled children, who are overweight and underexercised."[13]

Limited perspectives bring limited results.

Medicine should not be seen as a sporting event with the surgeon on one team, the therapist on the other and the patient as the ball. It takes the integration of good minds to bring better choices for scoliosis patients who have turned to us for help. It's only through this approach, using all the available tools, that we can prevent others from suffering the way Tom, Laura and many others have, as you'll read about next.

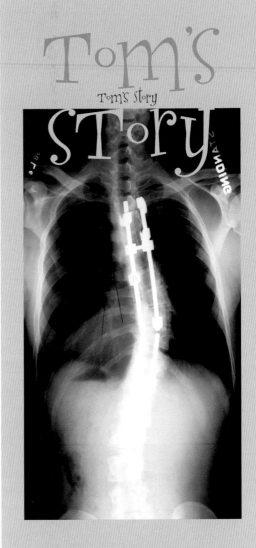

Tom's Story

"Please don't be offended if I don't believe you right off the bat. I've talked to so many medical people about this, it's hard to believe that anyone could get me to the point where I do not have to live with this pain. But I will do whatever you say."

This was one of the first things Tom said to me as he moved stiffly through my door. Tom told a sad, frustrating story of how he ended up with rods in his back to stop the pain of his scoliosis. Surgery was the usual procedure for people with spines as curved as his 56 degrees.

Prior to the surgery, he was told that after a month he would be feeling great. In a few weeks, he would graduate from college and turn 22 and be dropped from his parents' insurance. If the surgery was done then it would be paid for, and he'd be out of pain. Otherwise, the $60,000 surgery would be his responsibility as a pre-existing condition.

He had the surgery. And now, at 30, was still not out of pain. The rods broke in his back. His only alternative for the pain was to have the rods removed. Another surgery, with no guarantees.

After exploring various medical avenues without finding relief, he created his own workout routine to get relief.

When I heard Tom's history I knew fairly quickly why he might have progressed with pain.

Tom hadn't been in pain when he was young and he'd been active in sports in high school. He'd played competitive soccer since 5th grade and had a large paper route which meant walking and lifting papers every morning for an hour, even carrying them on his head and shoulders. In college he stopped working out and did more hanging out. His new sport was partying and by his senior year he could not sit in class without pain.

About a year before we met, he'd started swimming and felt the first relief in years. An hour or two after a swim he felt less stress and less pain.

As we talked, Tom began to put the pieces together himself: if he'd kept moving and been active in sports, his story might have been different. Correct movement is critical for people with scoliosis.

I believed that as young and motivated as he was, it would not take much to get him feeling good. He was to continue swimming and see me once a week.

Since a good portion of his spine was fused, the top priority was to regain full natural range of motion in his hips and for his legs to be balanced correctly under his spine.

Tom's first step to pain freedom was relaxing unwanted holding patterns in his hips and back. These patterns developed because, for example, he was afraid to bend over at the hips because of back pain so he kept his body tense and held his back and hips in the wrong places.

When relaxed, his body's many receptor systems were able to get the messages they needed so he could move without pain. He was now adapting and coordinating his body in a natural way. His movement slowly began to change because he started to use his joints as they were designed to be used. He was gaining optimal posture and realized when he knew how to move he could bend at the hips without pain. Tom started to realize that he had been working to "hold" his body with tension in an effort to protect it when it really worked the opposite way. That is, when he relaxed, movements could be done correctly in accordance with his body's design.

Ten weeks later, Tom told me he now noticed that when he did have some pain (like the night he slept in his truck) it bothered him a little!

"I don't think people know what it is to never be out of pain. When you are standing and talking to someone, he has no idea that your back feels like someone has their fist in it all the time. No furniture is comfortable, so why sit down, or even lie on a bed? You only sleep when you are so tired you can't stand up any more."

Tom wants others in pain to know there are other choices besides surgery.

Laura's Story

Laura's Story

Laura likes challenges. It's just as well, because her whole life has been one of over-coming physical challenges.

The story began in 1970 when Laura was 12 and diagnosed with scoliosis. Even though she wasn't in pain, surgery was chosen as a preventive measure, and inserting Harrington rods (described earlier) was the new and fashionable surgery. She was first "prepared" for surgery by having her spine "straightened" on a torturous machine called "the rack," then she was immediately put in a cast for five days to force her spine into the desired position.

There were many complications to the subsequent surgery. She contracted blood poisoning and hepatitis from a transfusion.

Her body cast had to be removed because of massive swelling. Her incisions would not heal.

It was April 1971 before she was cast-free. Just as she was healing, in January 1972, the rods broke. More surgery was ordered, to correct the first surgery. This young teenager spent seven months in a full body cast and another four months in one where she could move her arms and legs. But this wasn't the end of her imprisonment in plaster.

For another seven months she was in a "turn-buckle cast." A buckle was turned to move her body to fit her spine. When her body and spine were finally in the desired position (after approximately six weeks), Laura underwent another surgery to fuse her spine a second time. She remained in the turn-buckle for another six months.

"I can still remember being on my stomach with my head facing straight ahead in a most abnormal position. My body was put in the strangest contortions."

So, four years, four surgeries, blood poisoning, hepatitis, the rack, and the turn-buckle cast later, she was able to have a normal life. Normal for 15 years, anyway, until she was diagnosed with breast cancer at 30.

She has been a cancer survivor for more than 10 years now. Laura feels the multitude of X-rays she had as a child contributed to the cancer, and research supports this thought.

Just to recap: At the age of 12, before all this medical intervention she had scoliosis but no pain. Following her spinal fusion, Laura led a relatively pain-free life until problems surfaced again at 35 when lower back pain surfaced and advanced to the point where she could no longer stand at the sink to wash dishes.

Laura had been very active for 15 years doing aerobics and competing in high level tennis daily. She loved being active. The pain frightened her. She went to an orthopedic specialist on spines who referred her to a physical therapist. Despite faithfully carrying out her exercises, she felt worse. A second specialist told her that steroid injections would be the only option, but they would only give her pain relief for a limited number of years.

An MRI brought more bad news. It showed degenerative changes below the fusion, a common side effect of an extensive spinal fusion. It also showed one vertebra sliding off another (spondylolythesis at L4 and L5) and a narrowing of the hole where the spinal nerve exits (encroachment of a foramen on the right at the same level).

A second specialist told her she could only swim and ride a recumbent bike. He advised her not to play tennis, run or do aerobics and warned that down the road she would be confined to a wheelchair.

"My life was wiped away right in front of my eyes," she told me when we met shortly after this.

I first had her rest her head on her arms against a wall, and relax her entire pelvic area. This forced her to stop holding her stomach tightly. She said that was the moment she felt her first relief.

Laura did the *Body Balance* moves faithfully. She's been out of pain for five years now, and she's as active as she chooses. She plays tennis up to twice a day. She carries her two- and five-year-old sons up and down three flights of stairs daily. She was delighted to compete in a 350-mile, four-day Texas AIDS ride with her 17-year-old daughter, and is currently training to power walk a marathon.

It is obvious this woman recycled her pain into something extraordinary.

These days, Laura's challenges are confined to her work among the less fortunate, or on the tennis court, or keeping up with her kids. And yes, she feels fine standing at the sink now—when she has time!

Today the routine of treating scoliosis is more sophisticated. Neither the turn-buckle cast nor racking is used. But young people with scoliosis still often spend many years in braces and undergo multiple surgeries, as did Laura.

Perhaps more people can physically relate to these next stories. Clarice and Ira just wanted to enjoy their lives, and not knowing how to move their bodies was holding them back.

Some people are so motivated that their transformation is inspirational.

Clarice is such a person. She did not have the time or the money to fiddle around with back pain. It had already stopped her from driving, doing housework and even teaching her children in home school.

Clarice came to me in January 1996 with a history of back and leg pain. Her problems began in June 1995, and now, she said, "I am unable to do anything." She was 45 years old, had seen a slew of medical professionals but had had no relief.

She had sciatic-related pain and her pelvis had rotated to create a "flat back," which means no lumbar curve. This caused her center of gravity to be off and for most of her weight to be thrown over her toes. This put an undue amount of stress on segments of her spine, especially the lower back area. It also caused her to have an abnormal forward head posture. She had scoliosis as well. While it was not severe, it contributed to her problems. Many of her joints had lost their sense of "awareness" of how to move. Because of that, she had a very difficult time learning to move naturally again.

The first picture was taken on February 2, 1996, on her first visit. The second picture was taken three weeks later after two visits and six *Body Balance* classes. Notice how by

First Visit   3 Weeks Later

simply moving her pelvis and leg bones into a position that carried more weight properly, her excessive spinal curves decreased. This chain of responses along the spine helps prove the body was meant to be that way. Once her lower body alignment changed, her forward head position naturally decreased. Her weight is now distributed more in her heels which allows healthier movement. She was able to drive and do basic activities again. Within the next couple of months she was functioning normally. Christmas 1998, I received a letter from Clarice telling me how good she felt. She was one inch taller than at her last doctor's visit in June! She said she could now garden, cook, drive and do anything she wanted. "Good health," she said, "was like money in the bank."

Clarice's Story

Ira's Story

Getting things right is what keeps Ira's life going smoothly. He likes being able to go from work to the golf course without a glitch. In 1985, Ira had a big glitch that sent his back into spasms and him to the hospital. These back-spasm episodes did not stop although their severity lessened, and after a few years he was just tired of it.

Traditional physical therapy had not helped. In January 1997, Ira joined my *Body Balance* classes. He practiced diligently. Like any golfer, he knew all about practice and wanted a good back to go with his golf swing. Ira does not have a lot of flexibility but this isn't as vital as many golfers think. However, some of his joints did need to move more, and as Ira said: "You taught me how to bend from the hips and keep the 'v' in my back" (the natural lumbar arch). It took some time to get back the natural motion in his hips and spine so he could bend with ease, but he claims this knowledge, and correct postures, contributed to the success he still has three years later.

As a golfer, Ira knows that a good setup generates a good swing. Now he also knows a good *Body Balance* posture setup gives him a great pain-free swing.

Ira incorporates *Body Balance* into his way of life. He spends much time at his desk, but his new way of living at his desk helps his performance on the golf course. Ira plans to play golf longer and if he continues to integrate *Body Balance* principles, he knows he can only get better.

George Lynch's Story

Body Balance therapy "started out being weird," says pro basketball player George Lynch, "it's really unusual but it works!"

Lynch, the power forward for the Philadelphia 76ers, had a golden career—until he was hurt. As a University of North Carolina player, he went to two NCAA Final Fours and one National Championship. His experience illustrates that how you train will greatly influence how you play your sport.

While strong trunk muscles are vital, your abdominals are just one portion of these powerful movers. George Lynch discovered that doing abdominal exercises the old fashioned way took him off the court and almost out of the NBA.

He also found that many of the stretches and moves that were supposed to keep him well were actually stressing his tissue and setting him up for injuries.

George came to me after being diagnosed and treated by many "experts" after being injured playing in Vancouver. He suffered pain in his lower abdominal region whenever he tried the slightest acceleration down the court. It got to the point that when he would rotate and take a few steps forward, he had to stop because he was in so much pain.

The first thing I did with George was stop him from doing what I call "mindless movements." These included abdominal crunches and groin stretches that continued to place his pelvis in an incorrect position. Remember, tissue adapts as you train it and someone who is 6 ft. 7 in. must train his trunk muscles in the manner they will be used, otherwise he will be at a mechanical disadvantage, a bad career move.

So why do I call these moves "mindless"? Well, the power of this basketball player to get down the court depends on the position of his spine and pelvis so the right muscles can work. If he trains his abdominals with crunches, repeatedly, he will create muscle imbalances and bring his spine and pelvis out of the natural alignment he needs to play basketball.

If you do stomach crunches, you will be good at that one movement but that doesn't mean you will do better on the court. The trick with George was to show him that all of the fancy abdominal exercises he had been taught only aided him in those fancy exercises—not in sinking the basketball or weaving through the opposition down the court. I showed him another training formula that would help him improve his trunk muscles in a synergistic way.

Remember, abdominal muscles automatically work when you walk, bend and move during everyday life if you move in balance. *Body Balance* gave George a new understanding that what he did off the court influenced what he did on it.

Within three weeks George was back on the court and has now signed an even better contract with the 76ers.

George told me that my ideas reminded him of his great coach at North Carolina, Dean Smith. Smith believed training should be for the purpose of playing basketball and that weight training was overrated, according to George. When he first got into the NBA, their focus was on being big and strong and bulking up. Fortunately, he said, the NBA as a whole seems to be changing from strength-focused exercise to more finesse and speed-type training.

Barney's Story

Charm schools would love to hire Barney. This man is a natural who brings out the very best in those around him. When Barney hit a wall of injuries that kept him from his triathlon passion, he maintained his optimistic attitude and went hunting for the answer.

He was willing to try anything to get back on the track. He looked under every rock in town. His resume of therapeutic attempts over a four-year period would make a headhunter's head swim!

Barney's problems began when he was actively competing in triathlons. It started with pain in his right knee but he decided to work through the pain. Soon he was sidelined with shoulder, back, and knee pains that could not be diagnosed.

Whenever he tried to do more than his daily activities, pain struck. Even those long walks in airports set him back for days at a time. His recovery from training took longer and longer until, at 37, he called it quits. He had to throw triathlons in the trash because they were trashing him.

In the four years since his triathlon days he had seen a rolfer, an osteopath, a chiropractor, three physical therapists, a massage therapist, an acupuncturist and an orthopedic surgeon who did arthroscopic surgery on his right knee. He saw a rheumatologist and a neurologist and was planning to go to the Mayo Clinic to see if it was a metabolic dysfunction that kept him out of commission.

When he sought my help he had no great expectations, just a little hope and a great attitude.

He told me that, as a child, he had broken his right leg in five places. He had the usual shoulder aches and pains from being on the high school swim team. In his 20s, his right knee gave him a little problem when he ran too far in old running shoes. All the problems were on his right side. I could see straight away that changes were needed in his standing posture.

When I asked him to stand and look straight ahead, I noticed that he was facing slightly to the right. Then, when he started the *Body Balance* moves, he was unable to create smooth movement and maintain a balanced pelvis while standing on his right leg. When he lay down, he lay at an angle but thought he was straight. He had limited range of motion in his left hip. His right leg rotated too far in and away from the centerline of his body when standing. (I see this often and call it "the wandering leg.")

"You're twisted!" I told Barney. He laughed and said "That's news?"

It wasn't long before he saw progress using *Body Balance* moves. However, he was hesitant about being too active too quickly, fearing that all his injuries would flare up again.

I gave him a game plan: Get back on the bike, focusing on spinning (little resistance) and the position of your body. Use your knowledge of *Body Balance* principles to maintain form as a guide to how long you ride. When you feel you lose the form, stop. When walking, use *Body Balance* principles and focus on the way you walk. Again, let form guide the length of your walk.

To his surprise, Barney could walk further, golf for days in a row, and get on his bike again with minimal discomfort. Previously, such activities would result in days of pain and taking anti-inflammatories. Now, even the minimal discomfort faded after a day or two.

"To this day, instead of having trouble walking through an airport, I'm up to running four miles," Barney told me.

I told him the more he moved now, the better he would be and encouraged him to get back to swimming.

"You were just a little twisted, Barney, but we got most of the twists out. So, get out more and see what happens," I encouraged him.

Barney was not sure because the last time he hit the water his shoulder hurt for days. He'd grown up swimming on the swim team and the 'fly was his stroke.

After a few miles in the water putting the 'fly to the test, Barney could see how the *Body Balance* moves worked in the water. The moves fit well with swimming because the whole body is used in the water.

Barney told me the "'fly is a dance with the water" and now he's dancing without pain. His twist is gone, his 'fly is awesome, and Barney's smiling now that he has found the right way to move.

The Seasoned Athletes

Working with young athletes often produces quick results. There's not a lot of old patterning problems with their movements and their tissue is young and healthy. It is the seasoned athlete who gets my mental wheels moving.

When you hit 40, your tissue is usually not as pliable as it was when you were younger. There is not as quick a fix with those who may have pushed their bodies past the limit for a goal, or whose athletic careers started later in life. Their recovery from injuries can take more time and effort, but once they master the new moves they feel much "younger" than they had hoped in the beginning.

There is a huge, and growing, population now active in masters groups of sports ranging from swimmers to golfers and tennis players. Many other social players just want to keep playing their sport longer. These people will find the *Body Balance Method* particularly applicable.

Jim has been doing flip turns in pools since he was seven. He's probably spent more hours in water than most of us have slept. The work paid off in the 1976 Olympics in Montreal when he won three gold medals and a bronze for the USA.

Things went along fairly well until a car accident in 1983 left Jim with chronic lower back pain. The X-rays and MRIs after the accident showed his back already had suffered plenty of abuse and the accident was the last straw. He'd been left with a tremendous amount of abnormal breakdown for someone of his age and physical stature.

"It is common for freestyle and butterfly sprinters to develop these low back problems," according to Jim.

For five years he tried to eliminate his pain. First, he underwent the usual medical "cures" from physical rehabilitation to cortisone shots. When the chronic pain persisted, he ventured into nontraditional methods, such as acupuncture, which helped break the pain cycle, but he could not get back to competition-level swimming without pain.

A few years ago, Jim brought his 6 ft. 6 in. frame into my *Body Balance* world and has not been the same since.

"It really helped my back considerably when I learned the *Body Balance* moves. They gave me more flexibility and movement in my lower extremities. Getting the lower back right is so important since everything attaches there," Jim said.

Each day now, he rolls out of bed to do what he likes to call my *Body Balance* "passive stretches." His standing posture is

Jim's Story

Jim Montgomery's Story

much better. His buttocks are not as tucked and his pelvis and legs are in a better position for standing. This is particularly important since he stands on the deck and coaches for hours at a time.

Tissue responds to any type of conditioning program, meaning it adapts to whatever you do, good or bad. Swimmers at this high level have probably spent most of their lives horizontal: sleeping and swimming. It's no surprise they don't stand with ease. Learning to move "on land" is important for someone like Jim.

His career has continued since he learned these new moves, and he can do distance swimming without pain. For instance, at 40, he tackled a 10-mile crossing of the Maui Channel in Hawaii, a swim challenge that appeals to many master swimmers from around the world. Jim swam solo, competing against six-member teams. His one-man team came in 9th place out of 43 teams and he felt fine.

Jim now swims and stands without pain and attributes this to *Body Balance*. His back now knows the recipe for staying pain free.

Polio attacks the spinal cord segmentally, meaning it can skip a segment. Each segment innervates a specific muscle or muscles and when damaged at the cell level, this leads to either muscle weakness or paralysis.

Ph.D.s of Therapy

Few people have been through physical therapy for as many years as those affected by polio. Their physical therapists almost became family members!

Therefore it's no surprise to me that when I spend just one hour with a person who has had polio that they make profound changes because of their ability to tune in to how to move. They are always looking for a way to make movement flow a little easier and a little smoother.

For much of the first half of the 20th century, poliomyelitis was the disease that people in the United States, and around the world, most feared. It mainly attacked children. While some recovered, many were left with various degrees of paralysis, and many spent their lives on respirators.

Poliomyelitis is a viral infectious disease that destroys or weakens the nerve cells responsible for relaying motor impulses to the muscles, thus creating movement impairment.[14] Polio attacks the spinal cord segmentally, meaning it can skip a segment. Each segment innervates a specific muscle or muscles and when damaged at the cell level, this leads to either muscle weakness or paralysis. Our body's backup system is so powerful that half the relevant nerve cells must be destroyed for there to be clinical evidence of weakness in the muscle.[15]

Big muscles, such as the quadriceps, need several segments to stimulate them into action. The quads are innervated by three nerve roots (L2, L3 and L4). Unless all three levels had at least half their relevant nerve cells destroyed, there may be minimal to no weakness.

On the other hand, the muscle used to flex the foot towards the shin is mainly innervated by a single nerve root (L4). When polio affects L4, it causes the common foot drop seen in many polio victims.[16]

The Salk vaccine, then Sabin, virtually eliminated polio in developed countries, but sadly, it is still common in less developed countries.

However, in recent years, many people who contracted polio earlier in their lives are now being hit with post polio syndrome (PPS). Its most common symptom is fatigue, which affects 60 to 90 percent of post polios.[17] PPS may affect people who believed they had fully recovered from polio. Joint and muscle pain is another new symptom. There are some 640,000 "old polios" in the United States, about 20 to 40 percent of whom will develop PPS.

The exact reason for PPS is not completely understood.[18] However, it is important that muscles are not overused because this leads to pain, fatigue and weakness. Being skilled at moving well is an ideal way to get more "action" for less muscle use. That is why I see such wonderful changes in my polio patients. Their keen sense of awareness of movement helps them adopt the more efficient gait I teach with *Body Balance*. For them it means less work and less pain and fatigue. Give these people a little mechanical advantage and they will go far.

The muscle used to flex the foot towards the shin is mainly innervated by a single nerve root (L4). When polio affects L4, it causes the common foot drop seen in many polio victims.[16]

L4 Nerve Root

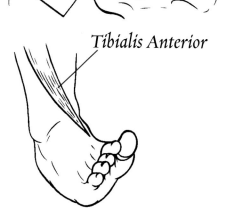

Tibialis Anterior

Valerie

Valerie (*above, left*) contracted polio when she was four. She believes water therapy early on could have been a big factor in helping her improve from being totally paralyzed to losing only selected muscles in her right leg. She wore leg braces and had a couple of surgeries as well as years of physical therapy.

After her last surgery, at 13, her parents took her for a checkup with the surgeon where they lived in Melbourne, Australia. They asked if anything further could be done. The surgeon said there was nothing clinically but then focused on her limp and added: "She could carry a big handbag!" It was a comment Valerie would never forget.

In the early 1990s, she was diagnosed with post polio syndrome and her brother arranged for her to see an American PPS

specialist who suggested she use a cane. He said there were two schools of thought. One, if you don't use your muscles, you'll lose them. Two, if you use them too much, you'll lose them. Neither answer seemed satisfactory.

The cane proved ideal. She could walk straighter and further. But there was a missing link: really knowing how to walk properly.

When I introduced Valerie to the *Body Balance* moves, she immediately saw how her cane "taught" her spine and hip muscles to move properly and get more glide in her gait.

"It was only when you showed me the *Body Balance* principles that I realized what power I could gain by altering the way I pushed off with my right leg, or swung my left," she told me.

"What was even more exciting was learning that walking wasn't just a waist-down activity. To be most efficient and graceful, all the moves I'd learned in *Body Balance* came into play."

Body Balance offered her another bonus. She has lived with low-grade back pain, but now sits much more comfortably. That's particularly important when she travels between the United States and Australia which can mean up to 29 hours on planes and at airports. She transforms an airline blanket or a pullover into an upper back rest when driving, flying or at her computer. This works much better by supporting her thoracic spine instead of the lumbar supports so commonly used.

Valerie admits to carrying too much junk in her handbag, but never uses it to disguise her limp.

Ann

Ann is a real Texas woman. She is not afraid to tell you just how she thinks or feels. What speaks loudly and clearly about Ann is the size of her hair, her heart and her red Cadillac. She gets the most out of life.

You won't find this lady at the back of the line in any event in life. She has had some hard hits, yet has come back stronger and ready for the next round. Polio was just a sidestep in life that did not interrupt this mom from raising her three children and being an executive's wife, with all its demands.

When she hit her late 50s, however, movement became a little more difficult and a few falls interrupted her stylish ways. Much of her energy went into worrying about not getting too tired mentally, because that would be her downfall, literally. The last spill split her head, necessitated 18 staples and sent her to a doctor who was not afraid of "alternative" ideas. That brought her to my doorstep.

Ann's list of hardships and recoveries let me know this lady was serious about "walking upright, right."

Like many who had had polio, she had gained weight because of the difficulty of increasing exercise without fatigue. That was not the case a year later. Ann found that with the *Body Balance* moves and a stylish cane, not only could she walk anywhere without fear of falling, but she'd also lost weight and moved with less fatigue.

She does swimming pool "jogging" for exercise and uses *Body Balance* moves to walk. I advised her to use a cane because I'd seen how it helped Valerie. It gave Ann confidence against falling. Then I showed her a more efficient position for her pelvis which allowed her to walk using less energy. Ann's gait is so smooth that most people would question whether she had polio. Her posture is close to optimal, which means she is going to work less to do whatever she decides to do.

The cane also helped stimulate the many receptor systems of the body that tell the trunk muscles what is normal movement, thus creating a universal change throughout her whole system. Even without the cane she is more upright because her body thinks that her moves are normal now and it responds to the new commands.

We eliminated the torso swing that many with polio have, and she looks great with 30 pounds gone as well. She and her family took a month-long trip to Africa, and she returned raving about her ability to walk as well as she did 20 years ago. This woman has found her fountain of youth.

Attitude Is The Key

I cannot take credit for these success stories because I didn't solve these problems. I have found that knowing how to move through life is not just physical, it also depends on attitude.

With a little humor and a lot of heart, each of the people here adopted and integrated the *Body Balance* principles they needed. They achieved results because of their desire to use what they already had to fix their problems.

I like to think of movement as a continuous enjoyment, no matter where you are, or what you are doing. And I think this attitude is catching on.

How do great athletes move when running, jumping, landing or turning? Look at the position of their spine and hips. Notice their ability to relax while doing complex moves.

1 Mary Loyd Ireland, M.D., Interviewed by Lisa Ann McCall, November 28, 2000, Dallas, Texas, personal,. p 31.

2 "Scoliosis: Diagnosing Girls at Risk," American Academy of Orthopaedic Surgeons, September 1, 1994, p. 1.

3 N. Schommer, *Stopping Scoliosis*, Avery Publishing Group, New York, 1991, p. 6.

4 Ibid., p. 7.

5 "Scoliosis: Diagnosing Girls at Risk," American Academy of Orthopaedic Surgeons, September 1, 1994, p. 2.

6 N. Schommer, op.cit., pp. 15, 57.

7 Ibid., p. 67.

8 T. Gavin, C.O., "Orthotic Treatment for Idiopathic Scoliosis: Current Concepts," *Backtalk*, published by the Scoliosis Association.

9 N. Schommer, *Stopping Scoliosis*, Avery Publishing Group, New York, 1991, pp. 68-70.

10 "Scoliosis: Diagnosing Girls at Risk," American Academy of Orthopaedic Surgeons, September 1, 1994, p. 2.

11 N. Schommer, *Stopping Scoliosis*, Avery Publishing Group, New York, 1991, pp. 32-33.

12 Ibid., pp. 36-37.

13 Ibid., pp. 37-38.

14 Frequently Asked Questions, www.post-polio.org, Gazette International Networking Institute (GINI), coordinator of International Polio Network (IPN).

15 S. Hoppenfeld, *Orthopaedic Neurology*, Lipponcott-Raven Publishers, Philadelphia, 1997, p. 73.

16 Ibid., p. 73.

17 www.post-polio.org, op. cit., p. 1.

18 Ibid., p. 2.

CHAPTER VI
ENJOY MOVEMENT

When left undeveloped, the senses give us chaotic information: an untrained body moves in random clumsy ways, an insensitive eye presents ugly or uninteresting sights, the unmusical ear mainly hears jarring noises, the coarse palate knows only insipid tastes. If the functions of the body are left to atrophy, the quality of life becomes merely adequate, and for some even dismal. But if one takes control of what the body can do, and learns to impose order on physical sensations, entropy yields to a sense of enjoyable harmony in consciousness.[1]

Flamenco dancing, like many ethnic dance forms, maintains the beauty of a natural balanced body. A long open body with the neck strong and in line with the rest of the spine. No tucking of the buttocks, sucking in the stomach or lifting of the chest.

CHAPTER VI

HOW TO ENJOY MOVEMENT

"Under the newer concept of disease, tissue damage is seen as more the result of normal bodily processes gone awry or disrupted than it is the dirty work of microbes or other external culprits." [2]

Dr. Blair Justice

The McCall Body Balance Method helps you understand the mind-body connection that allows you to think your way to good health.

The McCall Body Balance Method is not just about moving well but is a way to balance your body and mind so you are happier and healthier moment by moment.

Your body is here to help—when you know its language. This chapter explains how your body communicates and how you can learn to listen. It costs nothing but your attention. As well as talking about the mind's influence on your health, this chapter illustrates why you need to make better use of what you already have: your senses, your breathing, relaxation and even pain.

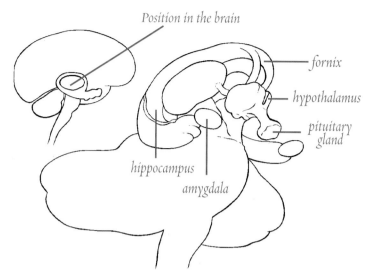

Position in the brain

fornix

hypothalamus

pituitary gland

hippocampus

amygdala

THE LIMBIC SYSTEM

Chemically Dependent

In 1977, Roger Guillemin and Andrew Schally won a Nobel Prize for demonstrating how the brain uses chemical messengers to give orders to the body.[3] By understanding the roles of these messenger molecules, you'll see the connection between your mind and your health.

The limbic system is the part of our brain that links our emotions, behavior and motivation.[4] Without it we could not adjust to external changes and would be like cold-blooded animals unable to maintain a constant internal climate.[5] The headquarters of this system is a pea-sized structure called the hypothalamus. It's the "brain of the brain" and affects most things you do, from regulating your temperature and chemical balances to controlling your heart rate, your sleep, hormones, sex, and emotions.[6]

"Headquarters" also controls the autonomic system, a part of the nervous system in charge of the involuntary functions. Thus, stimulating parts of the hypothalamus can cause antibodies to be released or the opposite, causing immune activity to be depressed. So you can think yourself into a healthy or unhealthy state.

The "brain of the brain"

The limbic system is the part of our brain that links our emotions, behavior and motivation.[4] Without it we could not adjust to external changes and would be like cold-blooded animals unable to maintain a constant internal climate.[5]

Fight or Flight, Rest and Repair

The sympathetic system is our Emergency Room. It controls emergency situations so we can cope with sudden changes . . .

The autonomic system is the part of your nervous system that controls those bodily functions that work without your conscious control.[7] It adjusts your body so you can stay comfortable and healthy in a changing environment by using two subsystems: the sympathetic and the parasympathetic nervous system.

The parasympathetic system is your Maintenance System. It maintains your heart rate, respiration and metabolism under normal conditions.[8]

The sympathetic system is your Emergency Room. It controls emergency situations so you can cope with sudden changes, such as during an athletic competition, combat, severe changes in temperature or blood loss.[9] It governs your fight or flight reaction to a stimulus.[10] As adrenaline and blood sugar rise, so do the amounts of other strong biochemicals in your body, preparing you for a physical response.[11] The sympathetic sirens blare and your body is ready for action. Heart rate increases, body temperature changes and pupils dilate so you can respond rapidly to a potentially disturbing external situation should you literally need to fight or flee.

But what if you are in an air-conditioned car with the stereo on and the disturbing situation is in your head? Your thoughts are focused on avoiding a traffic jam, anxious about the meeting you are late for, and still angry about the fight with your spouse. Those thoughts trigger your body to produce an excess of biochemicals that turn against you. Your body is unable to respond naturally, by fighting or fleeing, so is it any surprise we have road rage? All of these thoughts in early morning traffic influence your health.

However, it is your response to stress that is breaking down your immune system and creating the tension in your neck.[12] You sent your brain those "stress" messages. More worrying, if this stress lasts weeks or months or occurs frequently, it diverts energy for other body functions such as digestion, reproduction and growth. It also suppresses your immune system by releasing hormones that influence the white blood count, lowering your tolerance to disease.[13] This redistribution of resources may have pathological consequences.

Hans Seyle, an endocrinologist who spent more than 50 years researching stress and disease, used the following story to point out the real culprit.

If you pass a drunk who throws insults at you and you retaliate in a fighting manner, verbally or otherwise, you will discharge adrenaline that increases your blood pressure and pulse rate while your whole nervous system becomes alarmed and tense in anticipation of combat. If you happen to be at risk for coronary problems, the result may be a fatal heart attack. What, he asked, would have caused your death? The drunk? His insults? No, your death would have been caused by your own reaction.[14]

MASTERING THE MIND

So how do you stop your body from sending those nasty chemicals at an inappropriate moment? It's all about mastering our minds.

Yes, it is easier said than done for most of us. There are many Eastern gurus who have demonstrated mastering their minds by changing their heart rate and body temperature at will. You too can tap into that power with the control of your thoughts. This story from the book *Flow* exemplifies this concept well.

> Tension is often a habit. The good news is that relaxation, too, can be a habit.

"An acquaintance who worked in United States Air Force intelligence tells the story of a pilot who was imprisoned in North Vietnam for many years, and lost 80 pounds and much of his health in a jungle camp. When he was released, one of the first things he asked for was to play a game of golf. To the great astonishment of his fellow officers he played a superb game, despite his emaciated condition. To their inquiries he replied that every day of his imprisonment he imagined himself playing eighteen holes, carefully choosing his clubs and approach and systematically varying the course. This discipline not only helped preserve his sanity, but apparently also kept his physical skills well honed."[15]

> It is against our physical nature to be habitually in a non-relaxed state.

Just as lifting weights strengthens your skeletal muscles, you can literally work the muscle of the brain to keep mentally sharper longer. The mind is a wonderful resource. The secret is in how you use it.

Relaxation: The unknown friend

You should put learning to relax on your "to do" list and make it a priority. If you are not in a relaxed state, unwanted changes are being made throughout your body. By repeating anything, it becomes a habit, a way of life.

We are wired in such a way that specific patterns of neural connections in the brain create habit patterns. Thus, a movement that once had to be thought out becomes automatic. Tension is often a habit. The good news is that relaxation, too, can be a habit.

It is against our physical nature to be in a non-relaxed state on a habitual basis. Yet we "hold" our bodies in ways that create patterns that prevent freedom of movement. We should get better acquainted with relaxation, our "unknown friend" which can unlock awkward holding patterns.

Increased physical awareness helps us relax by making us "hear" when our body is tense. The *Body Balance* method helps us gain a relaxed state through greater awareness. Whether we are reading a book, or dancing the cha-cha, we can enjoy the experience more if we are physically more aware.

Consciously training yourself to be aware of your body becomes a habit. For example, share the next few moments with me.

As I sit here and type, I am constantly reminded how comfortable it is to be here. I may touch my shoulder and feel it release and drop slightly. Using this subtle reminder, my arms stay relaxed and my breath is comfortable. I am building an understanding that being comfortable is important in each moment of my life. I think about how I don't have to constantly shift in my chair searching for a restful position. I pause to become involved in these small moments. It becomes a requirement because I am tuned in to my inner world. I become conditioned and in a way spoiled (as I should be), because this is the way life was meant to be lived, taking each moment and enjoying it.

*You need the right kind
of touch in order to grow,
even more than vitamins.*

Duke University pharmacologist
Saul Schanberg

Touching AnoTHer CulTure

Lack of touch is harmful because the human skin needs constant stimulus to keep the body healthy and alive.

Notice that I use my sense of touch (touching my shoulder) to help me relax. Resources to help you relax are literally at your fingertips.

Our five senses are constantly available so that we can take in the wonders of the world. If we use what we already have, we don't just go through a day, but actually enjoy it.

Some senses are more developed than others. Our sight and hearing can cloud our too-busy minds and hinder our awareness of other senses that are lying underdeveloped but ready to enlighten us when given the chance.

The "science of the senses" is what I have experimented with in my work for years. Why did my touching their necks create better motor responses from my patients than instructing them through verbal and visual cues? Waking up the joints to gain the desired movement greatly depends on the touch of the hand.

Touching Another Culture

When I was 14 years old, we took a family trip to San Miguel de Allende, Mexico. I was at that delicate age of being into what was "cool" in manner and dress. One evening, my sister and I walked down to the crowded square, joining the locals in their customary evening walk. To my surprise, the women were walking arm in arm as they shopped and socialized. So, my sister and I did the same. I remember the experience well. I knew it would be very "uncool" back home, yet I really enjoyed it. It made me feel wonderfully connected to her and safe in a foreign place. It had such a profound effect on me that I reflect on it to this day.

We Americans generally greet each other formally and with little or no touching. Lack of touch is harmful because the human skin needs constant stimulus to keep the body healthy and alive. We are concerned about having "our space" more than we are about placing an arm around a friend to show we care. We do not flow with the comfort of touching a shoulder, or kissing a cheek, as people in other countries in the world do naturally.

I have been teaching movement classes for years and see how our senses of sight and hearing are disproportionately stronger than our sense of touch. We are "in our heads" too much and don't feel life enough. There are millions of receptors in the skin and 2,000 in just one fingertip.[16] We can pick up much more information if we use this underused resource in understanding movement.[17] The connection between relaxation and touch is very strong, so is it any wonder that we are so uptight?

We do not stop needing touch in our emotional and cognitive development. None of our senses "grows up" to where we do not need them anymore. The receptors of the skin are one of the most important means of creating a communication of knowledge that directly relates to movement.

There are millions of receptors in the skin and 2,000 in just one fingertip.[16]

Touch Saves Lives
Science shows how touch can save and improve lives.

■ Tiffany Field, child psychologist and director of the University of Miami's Touch Research Institute, researched premature infants. These babies were massaged for 15 minutes three times a day and gained weight 47 percent faster than preemies given the standard care. The massaged preemies did not eat more but absorbed the food better or processed it more efficiently.

■ Field found that the massaged preemies were also more alert and aware of their surroundings, and their sleep was more restorative.[18] This effect continued. The savings alone should convince anyone that this was a good investment on many levels. The massaged preemies were discharged six days earlier, saving $10,000 per child.[19] When the babies were tested eight months later, they scored better on mental and motor skills than the non-massaged babies.[20]

■ Cradling a baby gives more than a wonderful feeling; scientists have discovered it is vital for a newborn's cognitive and emotional development.[21]

■ Research has shown massage helps ease the pain of burn patients and boosts AIDS patients' immune systems.[22]

■ Touch has kept many elderly people enjoying life, according to a study of 60-year-old-plus volunteers who were trained to give massages to preschoolers and toddlers and who received massages themselves. Giving the massages proved to be even more beneficial than receiving them–the volunteers exhibited less depression and loneliness and their stress hormone levels decreased. Even their daily lives changed. They drank less coffee, had fewer doctor's visits and became more socially interactive.[23] "Massage is medicine" is illustrated well by these examples.[24]

The less we touch, the more we are isolated from ourselves as well as from each other.

The power of this simple tool is underdeveloped and often misunderstood. There is nothing that causes a more dynamic, all-encompassing and speedy change than the touch of the hand. It paints a picture on far more levels than words, teaches us how to move well and connects people. In my classes, I am strict about few things. My rules are:

1. You must have fun.
2. You must view the body as interesting on the smallest level.
3. You must, must, must touch your body to understand movement, relaxation and how your body works.

We are touchy in our culture about this touch thing. Yet it is an easy and healthy bridge between people and our ability to relax and change our own body at any moment. The less we touch, the more we are isolated from ourselves as well as each other.

When practicing the *Body Balance* moves in Chapter 4 be sure to use your hands so you can get in touch with relaxation and how your body moves.

We do not stop needing touch in our emotional and cognitive development. None of our senses "grows up" to where we do not need it anymore. The receptors of the skin are one of the most important means of creating a communication of knowledge that directly relates to movement.

Emmaline Barker (left) is 88 years old. She became a massage therapist at the young age of 63. Emmaline states "retirement was not for me," and has enjoyed her third career now for 25 years.

In the excitement of getting his picture taken, Clinton, her "patient," had trouble lying still. But immediately Emmaline's touch calmed him.

The Other Side of Touch

While the sense of touch is a form of contact with ourselves and the outside world, kinesthesia is touch from the inside out; it's how you feel when you move. It could be thought of as the flip side of touch.

What do a golfer, a dancer, an archer and an ice skater have in common that gives them the ability to perform precisely? The feel of the move. Kinesthesia is your ability to be aware of where your joints are as you move. You gather the information through your proprioceptors.[25] Proprioception is your body's way of absorbing information through your skin, muscles, joints and surrounding tissue through various sensory receptors.

Kinesthesia is the feel you are looking for when swinging the golf club or landing on the ice. This sense of feel is vital to let you swim the butterfly without drinking the pool. It is also something we have lost due to our limited daily movement.

What do a golfer, a dancer, an archer and an ice skater have in common that gives them the ability to perform precisely? The feel of the move.

Children are still involved with all their senses in a relaxed state so they can stand, bend and move with strength, ease and purpose. They don't need to hold their body tightly; they are in touch with how to use it and how it interrelates with the environment.

Most top athletes still have this feel of movement. To regain it, realize it is how you move. The smallest movement can make the biggest difference. The way you carefully place your hand on the golf club or how your foot knows where to land when running the race depends on your gaining the feel that begins with your everyday movements.

We do not appreciate fully that we are designed to move without breakdown, without the help of knee braces, expensive running shoes and other orthopedic intervention. In the 1960 Rome Olympics, Abebe Bikila of Ethiopia won the marathon running in his bare feet.[26] *

It all starts at the joint level. When an athlete pushes her body to excel at a higher level she must stay within the parameters of a relaxed form to maintain the feel so the skill of the move can be executed at the highest level. But it all depends on having the physical "know how" day in and day out.

Breathing: The Cheapest Health Food

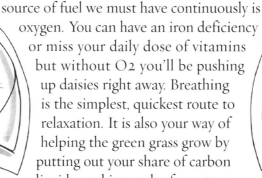

It seems self-evident to say "keep breathing" but many people just do not breathe properly. The one source of fuel we must have continuously is oxygen. You can have an iron deficiency or miss your daily dose of vitamins but without O2 you'll be pushing up daisies right away. Breathing is the simplest, quickest route to relaxation. It is also your way of helping the green grass grow by putting out your share of carbon dioxide, making each of us a true environmentalist!

INSPIRATION

EXPIRATION

How the Breath Works With the Body

When we do something well, it adds depth to the experience. Although we breathe, it is rarely a rich experience because it is influenced by the stress and anxiety of life. You can transform a stressful meeting into a walk in the park by bringing your mind to your breath and clicking into a relaxed state. When you focus on breathing, your body responds in ways that greatly change your physiological state.[27]

How we breathe can influence the quality of our breath.

The distribution of oxygen depends on the mechanics of breathing. The diaphragm (which separates the chest from the abdominal cavity) is a tough, flat sheet of muscle.[28] The chest cavity is like a cylinder that can expand in three ways, extending the diaphragmatic floor downward, the thoracic walls outward or the cylinder upward—all needed for the ideal complete breath. This is diaphragmatic breathing and physiologically it is the most efficient.[29] As the diaphragm moves down, it increases the chest cavity and, if the abdominal wall is relaxed, it moves outward as well.[30]

When we hold our abdominals "in" as the fashion magazines tell us, we may have a flat stomach at the expense of our breath. Flattening your stomach does not allow the air to go down, as designed. This form of breathing is more work because we use muscles in the chest not designed for that purpose. When we breathe correctly, we influence the autonomic system, creating the rest and repair response of the parasympathetic system that everyone needs daily.[31]

Math has its prime numbers; languages build on vowels and their sounds. Without either there is neither equation nor word. Without the kinesthetic sense, movement is awkward and clumsy, and may cause injury.

Pause for 30 seconds now

Do this experiment standing on an elevator, in the grocery line or right now as you sit and read.

Allow yourself to focus on four diaphragmatic breaths (doing it with your eyes closed can enhance the experience).

Relax your belly, back and pelvic floor.

As you inhale, take a "normal" breath but direct the air down toward your lower rib area, allowing the air to fill the entire space while letting your belly expand.

Exhale consciously, expelling all air. Notice if you feel more relaxed.

Breathing is one of the easiest ways to get in touch with the moment by simply noticing whether you are limiting your breath, or breathing naturally throughout the day.

Simplicity

By now, perhaps you have thought how simple this formula seems: breathe, touch, relax, enjoy movement. There's much to be gained from simplicity itself. It is the act of stripping away the paraphernalia of life and enjoying life's essence. It's about enjoying the movement in the moment which comes about by being so involved in the moment that there's simply no room for anything else.

I believe that one of the reasons golf is so addictive is that it takes many levels of involvement. Anyone who has picked up a golf club will admit that it is never without effort or frustration, yet people keep coming back for more.

Ask people if they enjoy walking on a treadmill and they will usually say no. They need a magazine to get through it. Walking, under these circumstances, is less mental work than golf, yet is also less satisfying. Being involved in whatever it is you are doing is the consistent secret to happiness.

Happiness lies in the simpler things, not the million-dollar toys. That is why walking in and of itself can be an enjoyment.

The simplest movements have the potential to bring the greatest pleasure. For example, kicking. That seems fairly simple. As an adult it may seem rather dull to kick a ball around, yet this simple act that we learn as a child is what the World Cup is all about. Place kickers do it, drill teams in Texas live for kicking, and the Rockettes travel all over the world being paid to do it. Just kicking. *Body Balance* shows how to break down your daily moves into smaller pieces, then reassemble them to make a smoother, more skilled, colorful and enjoyable movement.

Again, it is the how, the quality of the act, the mental involvement and the preparation that make the simplest acts monumental experiences. Kicking, throwing, walking: all are nothing without an awareness from the inside out.[32]

Two forces that drive human behavior and dominate our senses are pain and pleasure.[33] Sometimes it's the contrast of pain with pleasure that alerts us to appreciate a painless moment. Pain can literally suppress our immune system.[34]

Pain, defined by the International Association for the Study of Pain, is "an unpleasant emotional disorder evoked by sufficient activity of the nociceptive [pain] receptor system."[35] It is a subjective interpretation of a signal.

Pain comes from stimulating nerve fibers that are woven through tissue as well as embedded in blood vessels.[36] With increased anxiety, a hormone, norepinephrine, is released, and this causes vessel walls to constrict, which in turn can activate the pain receptor system.[37]

If this chapter is about using what we already have, how does that relate to pain? On the one hand, understanding pain is useful. Knowing our emotional state has the ability to create physiological changes that trigger the pain receptor to fire means that we can work on that.

Of even more use is the vast store of natural drugs in our brain. Science first discovered this in the mid-1970s and is con-

Pain

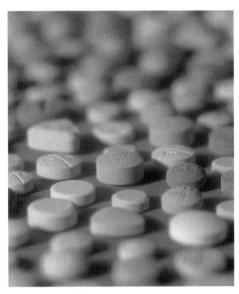

The Brain's Medicine Cabinet

stantly finding more "products" in stock. Here's a glimpse into the brain's medicine cabinet and how to use it to get relief from pain and anxiety.

There is the analgesics shelf. We often take medications like morphine for pain, but our bodies produce natural transmitters that can bring about the same effects.[38] These natural opiates are a variety of endorphins and enkephalins. The endorphins are the more potent of the two and are located in the brain as well as other areas of the body such as the intestinal tract and adrenal glands.[39] They can be activated by continuous vigorous activity if sustained for 15 to 20 minutes.[40]

Other messenger molecules participate in the body's analgesic experience. Norepinephrine, seratonin, substance P and dopamine are also on this shelf to help you cope with pain.[41]

We have neurons that secrete valium-like chemicals to calm us. Related to the endorphins, these transmitters are called benzodiazepines. They quiet anxiety, relax muscles and help induce sleep.[42] Research suggests we have the built-in potential to deal with both pain and anxiety by tapping

I studied the personality of pain and realized it is misdirected energy. Pain is a way your mind and body tells you things are not right.

"We create the pain and the suffering and the beauty in this world."

Tracy Chapman

these internal systems.[43] So what can help us get our own drug rehab program going?

Pharmacologist Avram Goldstein researched ways to stimulate good feelings. He surveyed 249 people, asking what gave them the thrills and chills, goose bumps, tingling feelings and intense sensations.

Music ranked first, followed by scenes from a ballet, movie, a play or book. Next came great beauty in art or nature, then touching, nostalgic moments, and sexual activity.[44] The simple act of laughing releases endorphins as well.

We need to begin to change our state of mind in order to break away from pain. If you continue to grin and bear it and think "no pain, no gain," you will end up in a pain cycle.

A pain cycle is trauma that leads to muscle guarding that leads to increased pain that further increases muscle guarding. Because of the cycle, tissue keeps being damaged as long as we continue with the incorrect movements. Our mental response or reaction to this cycle can make it better or worse. There is no one answer to rid the body from pain but there are many tools that can begin the process of helping to break the cycle. Pain does not warn us that something is going to go wrong but that trauma has already occurred. Pain is in your head as well as in your back, hip, neck or shoulder, and it is very real.

Life should not be about fixing your problems. Enjoy living. Whatever you do has the potential to be fun. It's all about your perspective.

TRAUMA

MUSCLE GUARDING

PAIN INCREASES

MORE MUSCLE GUARDING

WASTE PRODUCTS RETAINED IN TISSUE

TISSUE WITHOUT HEALTHY BLOOD FLOW

How we think and act will greatly influence our painful experiences.

Pain can invade the very fabric of a life. I have seen it drive people into despair and break them emotionally. While the sense of pain is a perception and not something you can hold in your hand, it should be seen as a powerful force.

All too often, we "fuel the pain fire" by our approach to the problem. The health care provider must be clinically knowledgeable and apply the knowledge in a humanitarian way. How many times have you laughed in your doctor's office? How many times have you felt your medical practitioner listened empathetically to your viewpoint?

I focus on changing people's perception by changing their focus onto something they can do to get themselves out of pain. Instead of looking at pain just as an unmyelinated C or A-delta fiber as I was taught, I studied the personality of pain and realized it is misdirected energy. Pain is a way your mind and body tells you things are not right. This is where the Seven Road Rules come into play.

2 Recognize when you are tired. Tiredness can lead to tissue damage and sap your immune system. It can fog your memory and make you unprepared for daily stresses. Get more rest if needed. That may mean a catnap at lunch or a mental rest through visualization, prayer or meditation. Try just rolling into bed at night a little earlier than usual. Have fun doing it.

1 Learn to move naturally. This will begin to change your breathing patterns, stimulate the neurotransmitters in your brain, the receptors in your tissue and joints that will keep the pain fibers from firing, so you can be more comfortable and relaxed. Moving as you are designed to move gives you control. Have fun doing it.

7 Be consistent. Do what is doable for you. Practice and it will become automatic. Start with whatever is enough to get you interested and see changes; this will have a snowballing effect. Have fun doing it.

3 Drink more water. You are 80 percent water. Dehydration damages tissue. Being stressed in the fight or flight mode is a workout even if you do not sweat, so replace the dehydration with water. Have fun doing it.

4 Check out your diet. Read and get some basic understandings of what you like to eat and, in return, what likes you back by making you feel better. Physically, your body needs time to eat for healthy digestion, so don't always eat on the run. Have fun doing it.

7 Road Rules to a Healthier Life

5 Pick moves that feel good. Swim the side stroke, your own "made-up" stroke or play tango music and dance as if the crowds had never seen such moves. Get creative about what it is that moves you to move and redefine the word "exercise." Have fun doing it.

6 Pause for the cause. Rushing to save two minutes to get wherever you are going may be more of a sacrifice than it's worth. Before getting out of the car for that meeting, lean back and listen to a few more minutes of Mozart, knowing the moment is all yours. Have fun doing it.

Quieting the Mind

In your office chair, parked car or in a quiet place in your home, sit or recline comfortably (to learn how to sit or lie down comfortably, turn to Chapter 4).

- Close your eyes and take three diaphragmatic breaths as you mentally follow the air into your lungs and around your lower ribs and back out again.
- Allow your body to relax as you focus on these breaths.
- Then begin to count down from 50 to 1, visualizing each number with each breath cycle.
- Allow your body to release tension as the numbers get smaller; imagine your body sinking into the chair or floor.
- If your mind wanders, bring your focus back to the numbers.
- Use color or draw the numbers in your mind, whatever will help you stay involved with the countdown.
- Don't worry about how often your mind wanders; it is the training to bring it back and stay in the moment that creates the change.
- When you have finished counting, take a moment to compare how you feel now with how you felt before. Give yourself a moment before going back into the "other" world.

Over time you will be surprised how it becomes easier to quiet your mind. You can tap back into the quiet feeling wherever you happen to be.

ReLax

*Some time today think
about your breathing for
two minutes. Relax your
abdominal area focusing
on each breath, and any
area where you feel ten-
sion, breathe and relax
that area (approximately
20 breaths if you'd rather
count breaths).*

1 M. Csikszentmihalyi, *Flow*, Basic Books, New York, 1997, p. 91.

2 B. Justice, *Who Gets Sick*, Peak Press, New York, 2000, p. 32.

3 B. Justice, *Who Gets Sick*, Peak Press, New York, 1987, p. 53.

4 R. Ornstein and R. F. Thompson, *The Amazing Brain*, Houghton Mifflin Company, Boston, 1984, p. 27.

5 Ibid., p. 27. (E. Kandel, J. H. Schwartz, and T. M. Jessell, *Principles of Neural Science*. Appleton & Lange, Connecticut, 1991, p. 733.)

6 Ornstein, op.cit., p. 28.

7 *Taber's Cyclopedic Medical Dictionary*, p. 159.

8 E. Kandel, J. H. Schwartz, and T. M. Jessell, *Principles of Neural Science*, Appleton & Lange, Connecticut, 1991, p. 761.

9 Ibid., p. 761.

10 Ibid., p. 761.

11 B. Justice, *Who Gets Sick*, Peak Press, Putnam & Sons, New York, p. 32.

12 Ibid., p. 55.

13 Ibid., p.22.

14 B. Justice, second edition, op.cit., p. 55.

15 M. Csikszentmihalyi, *Flow*, Basic Books, New York, 1997, p. 91.

16 Richard Wolkomir, "Charting the Terrain of Touch," *Smithsonian*, June, 2000, p. 44.

17 G. H. Colt, "The Magic of Touch," *Life* magazine, August, 1997, p. 60.

18 "The Senses," *U.S. News and World Report*, August 1999, p. 5.

19 G. H. Colt, "The Magic of Touch," *Life* magazine, August, 1997, p. 62.

20 "The Senses," *U.S. News and World Report*, August 1999, p. 5.

21 Ibid., p. 2.

22 Ibid., p. 5.

23 G. H. Colt, "The Magic of Touch," *Life* magazine, August, 1997, p. 62.

24 Ibid., p. 57.

25 B. Brownstein and S. Bronner (eds.), *Functional Movement in Orthopaedic & Sports Physical Therapy: Evaluation, Treatment & Outcomes*, ed., Churchill Livingstone Inc., New York, 1997, p. 43.

26 * He also won the Tokyo Olympic marathon in 1964, that time in shoes. Sadly, he was crippled after a car accident in 1969 and died in 1973.

27 Dr. R. Vaughn, interviewed by Lisa Ann McCall, April 2000, Dallas, Texas, phone conversation.

28 S. Rama, R. Bullentine, and A. Hymes, *Science of Breath: A Practical Guide*, Himalayan International Institute, Pennsylvania, 1979, p. 36.

29 Ibid., p. 38.

30 Ibid., p. 38.

31 A. Weil, Breathing Tapes.

32 M. Csikszentmihalyi. *Flow*, Basic Books, New York, 1997, p. 97.

33 R. Ornstein and R. F. Thompson, *The Amazing Brain*, Houghton Mifflin Company, Boston, 1984, p. 85.

34 B. Justice, *Who Gets Sick*, Peak Press, Putnam & Sons, New York, p. 52.

35 Ola Grimsby, *Clinical & Scientific Rationale for Modern Manual Therapy*, *MT-1*, Independent Study Manual, San Diego, 1998, *Neurophysiology*, p. 5.

36 Ibid., p. 6.

37 E. Kandel, J. H. Schwartz, and T. M. Jessell, *Principles of Neural Science*, Appleton & Lange, Connecticut, 1991, pp. 14, 183-184, 390.

38 B. Justice, second edition, op.cit., p. 117.

39 Ibid., p. 118.

40 B. Wyke, Presentation at A.P.T.A. Conference, 1974.

41 B. Justice, second edition, op.cit., p. 118.

42 Ibid., p. 122.

43 B. Justice, *Who Gets Sick*, Peak Press, Putnam & Sons, New York, 1987, p. 123.

44 B. Justice, second edition, op.cit., p. 121.

ENDNOTE

Many of the quotes in this chapter are from the book *Flow* by M. Csikszentmihalyi, Social Psychology professor at the University of Chicago. His book was written on what brings happiness. His philosophy on movement coincides with mine.

CHAPTER VII

BODY BALANCE
& YOUR BODY PARTS

The Communication System

CHAPTER VII

HOW BODY BALANCE WORKS WITH YOUR BODY'S PARTS

The Communication System

Knowledge is power. If you understand the basic workings of the body from a movement perspective, you will understand how movement is the key to good health. The next two chapters explain some relevant workings of the body. To keep it simple, we're going to start from the inside and work outward.

This chapter starts from the core: your brain and spinal cord, examining how you receive information from your body, and how your response through movement influences the health of your tissue. This knowledge will also make the *Body Balance* moves easier to do because you will be able to think about the different parts when you move.

> Your body is like one big satellite dish. It picks up information through tiny sensory receptors located throughout your body.

The Superhighway

The human mind and body have a communication capacity far beyond our understanding. A network of nerves, our neurological system,

constantly updates us on our internal and external environment and enables us to respond quickly to life's physical and mental demands.

It does this virtually without our realizing it. Our brain, spinal cord and nerves at work are a virtual superhighway of information.

Your Body, the Satellite Dish

Your body is like one big satellite dish. It picks up information through tiny sensory receptors located throughout your body. These sensory receptors collect information from joints and tissue and send messages to our brain via our central nervous system. When our brain and spinal cord receive information, they send an appropriate command, via a "motor response," to muscles and glands so we can react and move with coordination and grace.[1]

Different receptors pick up different information.

1. Static Receptors

Have you ever ached when sitting for a long time? Perhaps you slump over a computer at work? Slumping creates tension on the joint capsules of your hip and spine. Receptors in these joint capsules signal the brain about your position and the response is a dull, aching pain. You try to sit up straight, but it wears you out. You slump again; more pain. You can sit without pain by moving your joints into more ideal positions.

Often when patients come to me in pain, I will carefully place them in a more natural, ideal position or posture. They suddenly feel comfortable and relaxed. The static receptors are now providing a new signal.[2] They tune in to a new body position. The abnormal posture created abnormal tension on tissue, but in the correct position, these receptors inhibit the pain and the body can relax.

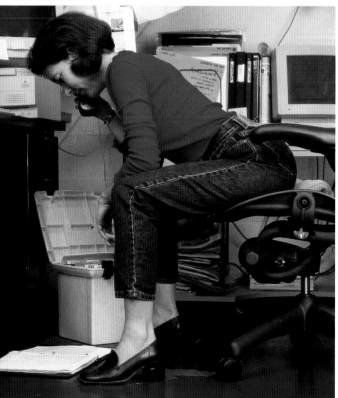

You can sit without pain by moving your joints into more ideal positions like Spectator Stretch shown here. See Chapter 4, Section 4.

2. Dynamic Receptors

Another type of receptor likes movement. These receptors activate when your body is in motion.[3] You'll find them in areas that need constant feedback for fine motor control, such as the lumbar spine, feet, hands and jaw (TMJ—temporomandibular joint.)[4]

These two groups—the static and dynamic receptors—work together. The static ones stabilize joints while the dynamic ones continue to fire throughout movement. In a golf swing, static receptors are responsible for the set up, while the dynamic receptors are responsible for the swing. This cooperative effort, without mental work, is called reflexogenic movement, which means the movement is automatic.

You'll find dynamic receptors in areas that need constant feedback for fine motor control, such as the lumbar spine, feet, hands and jaw.

3. A Pain Signal

We have all experienced the message that pain sends. It takes an abnormal level of mechanical, chemical or thermal irritants to trigger pain fibers.[5] When you sprain your ankle, for example, the excessive tension or stretch on the connective tissue creates pain because the pain fibers have been stimulated. You take the strain off your ankle, which eases the pain. Your ankle's temperature increases because the tissue damage causes inflammation. Chemical and thermal irritants build in the damaged tissue. Pain will continue until the inflammation is treated. Tissue damage can happen throughout the body with one or all of the irritants involved.

The neat thing is, you have the power to inhibit the pain fibers!

For example, if you hit your thumb with a hammer, you yell (or curse) then hold or rub your thumb. Why? Holding or rubbing sends a new and more desirable stimulus to the spinal cord which overrides the pain stimulus. This creates a change in the perception of pain because another, stronger message overrides it.

4. Receptors of the Skin

How is this ankle going to heal? Movement is the answer. With inflammation, there is a buildup of waste products and the trash needs to be taken out. This bad stuff (exudate) continues to do damage until there is movement to dissipate it and bring in important nutrients.

Movement also stimulates tissue to heal with minimal scarring. Mature scar tissue is unhealthy tissue void of blood vessels but housing healthy pain receptors.[6]

When people injure themselves in running or some other sport, they often tell me they "rest" their injury. While it is good to rest from the activity that created the damage, the tissue needs to have the correct movements for healing to occur.

The extraordinary fine tuning possible with skin receptors came home to me one night as I watched the Dallas Stars play ice hockey. I was amazed at how these guys with huge, cumbersome gloves on their hands had the finesse to feel the puck through the stick.

When you learn *Body Balance* moves, you get in touch with the many different types of receptors in your skin. By placing your hands on your body, you can harness your skin receptors in that area to help you relax. You also obtain a huge amount of information through the touch of your hand. "Hearing" a move is not as effective as feeling the move.

The extraordinary fine tuning possible with skin receptors came home to me one night as I watched the Dallas Stars play ice hockey. I was amazed at how these guys with huge, cumbersome gloves on their hands had the finesse to feel the puck through the stick. How could they even feel the stick? In spite of the gloves they could maintain the most delicate yet forceful grip and "know" quickly through their hands what to do. That is the beauty of the different receptor systems at work. Do not underestimate the power of your touch.

Receptors pick up the information given to them. If given dysfunctional information through dysfunctional movement, incorrect messages will be sent and received. Over time, the dysfunction can create pain and break down tissue. *The McCall Body Balance Method* reawakens a multitude of receptors throughout your body, giving you greater awareness of how to move.

Since most of your body's connective tissue is made up of water, drinking water helps keep your tissues healthy.

Cosmetic Turnover

Many women in the U.S. know collagen as the ingredient that fills the sagging spaces and creates beautiful bodies—with the help of "cosmetic" surgeons. There's another way to renew collagen. I long wondered why my patients began to look younger as they adopted *Body Balance* movement principles in their daily life. When correct physical stimulation (good stress) is applied to an area of the body by moving more in balance, your body rebuilds collagen. This can give you a healthier cosmetic appearance. Not only can your skin look more resilient but your shape gains more definition (buttocks lift instead of drop) and this gives an appearance of weight loss.

Collagen—It's Everywhere!

Collagen is the most fundamental component of the body's connective tissues.

As you read about joints, bones and muscles you will be able to see how collagen weaves itself into the very fiber of your body and how movement is vital to collagen's health and makeup.

Collagen is located in your body's connective tissue, tendons, ligaments, fibrous cartilage, fat, blood vessels, dermis, bone and joint capsule.[7]

Collagen is constantly being renewed and replaced so it must be fed and have the correct physical stimulation to continue this renewal process.[8] Appropriate movement is needed to regenerate our connective tissue. When movement is done incorrectly or not at all, the tissue becomes unhealthy, and undesirable changes occur. Since most of your body's connective tissue is made up of water, drinking water helps keep your tissues healthy.

Made to Move: The Synovial Joint

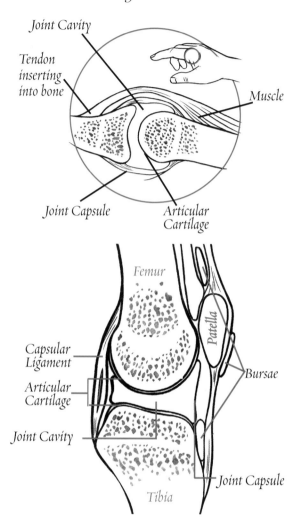

Joint Cavity

Tendon inserting into bone

Muscle

Joint Capsule

Articular Cartilage

Femur

Patella

Capsular Ligament

Articular Cartilage

Joint Cavity

Bursae

Joint Capsule

Tibia

The most common joint in the body is the synovial joint. The blood supply to the capsule and ligaments is limited, but they have rich nerve supplies. Large sensory fibers are sensitive to movement and position, and smaller sensory *pain* fibers are in the capsule, ligaments and blood vessels. These nerve endings are sensitive to twisting and stretching.[10]

Too much movement

At the other extreme, collagen can be damaged by overuse if it is stressed too much without adequate rest. This leads to Repetitive Stress Injuries (RSI). Carpal tunnel syndrome is a common injury of this type in people who work at computers.

RSI occurs when people sustain multiple micro-traumatic breakdowns to a particular ligament, tendon or functional system, e.g., the lower back or the wrist.[9] These breakdowns can contribute to your body's inability to successfully regenerate the damaged tissue.

Resting is one of the key ways to avoid, or to heal, this problem. Resting your tissue is more than getting a good night's sleep. It means resting tissue from unnecessary constant stress, such as being slumped at the computer. Appropriate rest with correct movement on a daily basis is vital for regeneration of tissue, and it speeds the healing process.

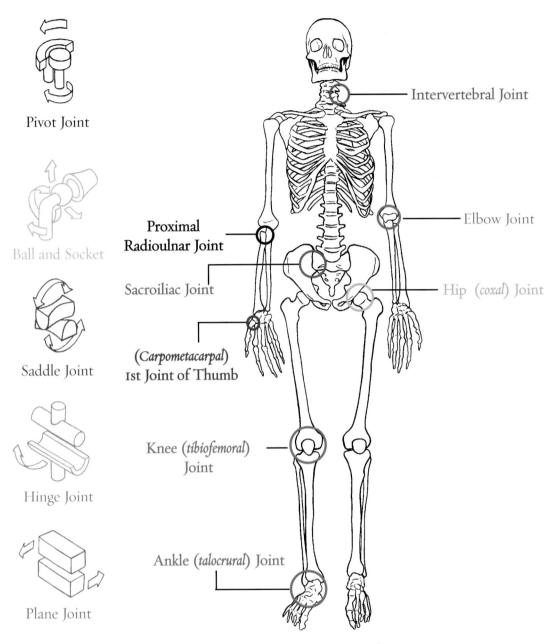

Pivot Joint

Ball and Socket

Saddle Joint

Hinge Joint

Plane Joint

Intervertebral Joint

Elbow Joint

Proximal Radioulnar Joint

Sacroiliac Joint

Hip (*coxal*) Joint

(*Carpometacarpal*) **1st Joint of Thumb**

Knee (*tibiofemoral*) Joint

Ankle (*talocrural*) Joint

Joints Move As Designed

Think of joints as hinges that vary according to their job. These specialized structures provide stability as well as mobility and play a role as shock absorbers for the skeletal system.[11] They influence motion in multiple ways. On the other hand, movement can help or hinder joints depending on the forces (both internal by muscles and external by gravity) placed on them.

The most mobile joints in the body, with the greatest impact on movement, are the synovial joints. There are five different types of synovial joints, each with a specific range and type of movement.[12]

There's the ball-and-socket joint, such as the hip, which allows movement in multiple directions. There are joints along the spine, called plane joints that allow gliding for short distances.[13] If the hip joint does not function through its full range of motion, the forces of movement will take the path of least resistance and the joints along the spine, or even the knee (a hinge joint), may move outside its natural range. This could, and often does, create damage to the surrounding tissue and the joint itself. Each joint stays healthy and happy if it moves as it was designed.

Lube Your Joints and Get More Mileage

We always have friction when we bear weight on our joints. We just want it to be normal. Abnormal friction can occur for multiple reasons including incorrect movement, lack of joint lubrication or not enough rest.[14] Synovial fluid is your body's natural lubricant that supplies nutrition to your joints and keeps friction down. Relate this idea of synovial fluid to the oil in your car. Without oil, friction in the engine increases, the engine parts heat up and expand, reducing smooth movement until the engine seizes. Friction can ultimately damage anything. An engine, your back and elbow or a pair of pliers can break at the hinge or junction of movement due to friction. Correct movement keeps the joint lubricated and the friction down.

Going against the grain: When movement is done incorrectly

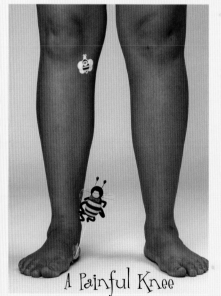

A Painful Knee

Fallen arches are created when the kneecaps turn in making the bee fall away from the line of the body.

If the hips are out of proper alignment, the thigh bone (femur) will be thrown into a different position relative to the lower leg (tibia). This can affect the feet. It can affect how the kneecap lines up in the femoral groove. If the kneecap is not in the groove, it will suffer unnecessary friction. That friction will create breakdown of bone where the two bones are rubbing together. Alternatively, if the hips are level and the leg bones in their proper place, there will be changes in the pressures in the kneecap, eliminating abnormal friction.

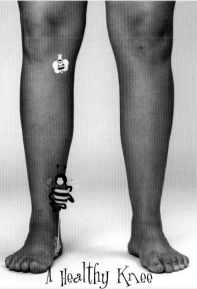

A Healthy Knee

When the thigh bones turn out and the feet stay planted an arch is created in the foot and the bee goes back in line with the body.

Worn-Out Joints

On the ends of all synovial joints there is a thin layer of cartilage. This articular cartilage (from the Latin word articulatus, to join together as a joint) has no nerve, blood or lymphatic supply.[15] When people say they do not have any cartilage in their knee or hip, it is this articular cartilage that is missing. This cartilage consists of collagen, so you already know what it needs to stay healthy.

Cartilage gets worn down by improper loading on a joint, too much joint load or direct trauma to the joint. Too much load on a joint can cause pain. The catch-all term for the problems affecting the cartilage is degenerative joint disease. Other problems, such as inflammation or the formation of small bones (osteophytes) or changes in the bone under the cartilage can then occur.[16]

I have helped people with advanced stages of joint degeneration decrease or eliminate their back pain and regain much of their functional trunk movement. One theory for this success could be that bringing the person's body alignment more in line with gravity creates a redistribution of forces through the skeletal structure and lessens the load on specific joints or specific areas within joints. We come back to the principle of distributing a load over the greatest surface area. The combination of more bone in line with gravity and greater surface area for pressure distribution means there is less demand on the joint tissues, thus less or no pain.

The Back Joints

The back is composed of two different types of joints:

The *annulus fibrosus*, is the outer portion of the disc located between the bones of the spine, the vertebrae. It is a fibrous joint.

The *facet joint* is the second type of joint in the spine. They directly articulate with the vertebra above and below.

Side View

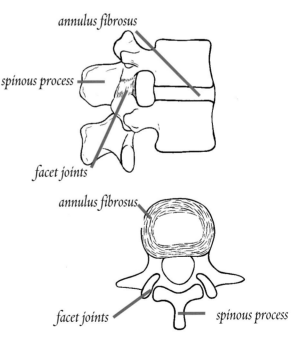

annulus fibrosus

spinous process

facet joints

annulus fibrosus

facet joints

spinous process

Top View

A Little Bit About Discs: Some Hurt; Some Don't

There is quite a controversy over disc lesions. When people hear their back pain diagnosed as a herniated (slipped), prolapsed or bulged disc, they often fall into a fatalistic tailspin out of proportion to the problem. The disc's primary function is carrying the weight of the torso and upper body. It also provides flexibility for the spinal column and cushioning.[17]

Research shows there are many pain-free people walking around with herniated discs. A study in the *New England Journal of Medicine* involved 67 people without symptoms of back pain. MRI (magnetic resonance imaging) examinations showed that 20 percent of the people under 60 years old had herniated discs. In people over 60, 36 percent had disc herniation. In another study of 41 women without symptoms, just over half had at least one bulge or herniation. This exam focused on the last three levels of the spine, which is the most common area for problems.[18]

Back pain involving disc dysfunction cannot be a simple diagnosis because many parts are involved when the back gets into trouble.

The disc itself can be creating the pain, but so can other parts. When an athlete hurts a knee and tears one of the main stabilizing ligaments, usually other supporting structures in the area will also be traumatized. Look at the back like any other joint and care for it in the same way.

An Older Disc

Diagnosing "disc degeneration or degradation" as the cause of pain is somewhat questionable.

The nerve fibers are located only in the outer third of the disc. Therefore, if the inner part is severely degenerated, you may not feel pain.[19]

On the other hand, a disc that looks relatively normal may have a small fissure or tear in the outer part that would create pain.[20]

The diagnosis of degenerative disc disease has a powerful negative effect on many people. It should be handled more intelligently by the medical community instead of promoting the assumption that the problem will only get worse or that when present it always causes pain.

annulus fibrosus

outer third

Internal Disc Disruption

The back has a similar structural makeup as the knee and hip. It does not need to be protected from movement but needs to move as it was designed.

Uncle Bob's Disc—overy

Rotation of the spine is good for your back. How can that be? You may ask. We all have been taught that rotation was what caused Uncle Bob's back to "go out" which left him laid up for days on pain medication.

Luckily for him it was not his disc! This poor anatomical part has received such a bad rap that now we do not let Uncle Bob lift anything because he might hurt his disc. It sounds more like a terminal illness that will somehow invade poor Uncle Bob's body if he moves at all.

Uncle Bob is probably hurting more from the argument he had with Aunt Betty. In reality, both his relationship with Aunt Betty and his posture were a mess. These caused tension which prompted muscle guarding in his back which led him to move off of the natural axis of his spine—all due to his condition before he moved. Now he's inactive and this helps trigger the pain receptors.

So there he was, in pain and out of action. He was told to have an MRI. This cost his insurance company $1,500 and showed that his problem was not his disc.

Now Uncle Bob does not know what happened, he just knows he has a "bad back" so he will do his sit-ups or press-ups every day (neither help tissue healing of the spine), stop playing golf and stop bending or

lifting because "it is bad for his back." He just hopes he does not have another episode.

What happened to Uncle Bob's back could have been a number of things because of the tight interrelationship of tissue, structure and function with other systems of the body.

What could help Uncle Bob are those very movements he wants to avoid along with a mental perspective of confidence and understanding of what is happening.

The *Body Balance* movements that could start him on his way to health include the floor movements that begin to introduce rotation from the hips to the lower-back spine (Rock and Roll, see Chapter 4). The "Rib Shift" exercise could help change the position of his mid-spine and allow the correct receptors to fire along his spine, which in turn would activate the correct muscle groups. All movements would have to be pain free to increase healing. Then postural treatments, in supported sitting and lying positions, would be vital to activate the postural muscles and prevent the pain and muscle spasms.

In this scenario, the mental relief and realization that with correct positioning he could start moving without pain teaches Uncle Bob to be independent. It gives him control over the problem. The "exercises" that brings him back to health are the very movements he wanted to avoid because he did not know how they work to generate the healing of healthy tissue. The result is that Uncle Bob would then have a healthy back because he retrained his postural muscles to move in the painless range. The proper knowledge would keep him fitter. He may even end up enjoying golf once he learns how correct Universal Movements stimulate healthy tissue in its ideal way, the natural way.

The back, like the knee or ankle joint, gets into trouble if it moves off of its natural axis. All the things that happened in the earlier example of ankle sprain swelling, soreness, inability to use it, etc., applies equally to your back. If your back joints move as they were designed to do they will stay healthy.

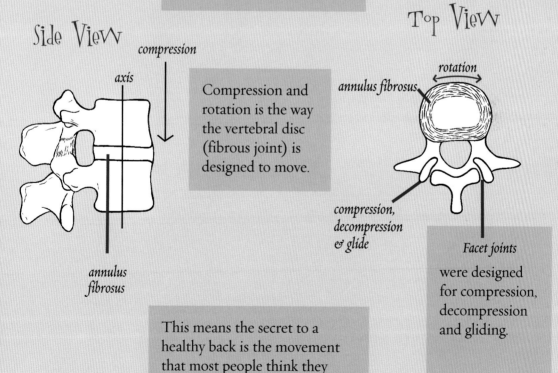

Side View

compression

axis

annulus fibrosus

Compression and rotation is the way the vertebral disc (fibrous joint) is designed to move.

Top View

rotation

annulus fibrosus

compression, decompression & glide

Facet joints were designed for compression, decompression and gliding.

This means the secret to a healthy back is the movement that most people think they should avoid: ROTATION of the spine.

Notice

Stand in front of your chair and sit down. Now close your eyes and go into the sitting position again. Notice how dependent you are on your joints to tell you how to do the same act without the help of your vision.

1 E. Kandel, J. H. Schwartz, and T. M. Jessell, *Principles of Neural Science*, Appleton & Lange, Connecticut, 1991, p. 22.

2 Wyke, 1979b; Freeman and Wyke, 1967, *Manual Medicine*, Thieme Medical Publishers, New York, 1990, p. 36.

3 Ibid., p. 36.

4 Grimsby, op.cit., p. 17.

5 D. Currier and R. Nelson, *Dynamics of Human Biologic Tissues*, F.A. Davis Company, Philadelphia, 1992, pp. 9, 10.

6 B. Parks, The myofibroblast anchoring strand–the fibronectin connection in wound healing and the possible loci of collagen fibril assembly. *J Trauma.* 1983;23(10):853-862.

7 N. Bogduk and L. Twomey, *Clinical Anatomy of the Lumbar Spine*, Churchill Livingstone, Melbourne, 1991, p. 153.18.

8 D. Currier and R. Nelson, op.cit., p. 32.

9 Bryan Williamson, P.T., MS, interviewed by Lisa Ann McCall, August 1999, Dallas, Texas, personal.

10 J. A. Gould and G. J. Davies (eds.), *Orthopaedic and Sports*, C.V. Mosby Company, Missouri, 1985, p. 105.

11 Ibid., p. 50.

12 *The Visual Dictionary of the Skeleton*, First American Edition, DK Publishing, New York, 1996, pp. 42-43.

13 Ibid.

14 Williamson, op. cit.

15 Saviol-Y. Woo and Buckwalter, Injury and Repair of the Musculoskeletal Soft Tissue, 1987 Workshop, American Academy of Orthopaedic Surgeons, pp. 428-429.

16 J. A. Gould and G. J. Davies (eds.), *Orthopaedic and Sports*, C.V. Mosby Company, Missouri, 1985, p. 114.

17 N. Bogduk and L. Twomey, op. cit., p. 20.

18 M. C. Jensen, M. N. Bryant-Zawadzki, N. Obuchowski, M. T. Modic, D. Malkasian, and J. S. Ross, "Magnetic Resonance Imaging of the Lumbar Spine in People Without Back Pain," *The New England Journal of Medicine*, July 14, 1994.

19 N. Bogduk and L. Twomey, *Clinical Anatomy of the Lumbar Spine*, Churchill Livingstone, Melbourne, 1991, p. 153.

20 Ibid.

CHAPTER VIII
BODY BALANCE
& YOUR BODY PARTS
The Action System

HOW BODY BALANCE WORKS WITH YOUR BODY'S PARTS

This chapter is about the roles muscles and bones play in reaching our ultimate body goal of moving in comfort.

The Action System

In Chapter 7 we examined what triggers our motion; now let's move further from the core and examine how we act out movement. This chapter is about the roles muscles and bones play in reaching our ultimate body goal of moving in comfort.

Have you ever wondered why repetition makes difficult tasks easier? Actions such as riding a bike or skiing down a mountain go from what is called a cortical level (the cortex of the brain or higher centers) to a less conscious level of the nervous system through repetition.[1] Thank goodness, or it would take hours to get dressed every morning!

Obviously, maintaining correct posture through movement does not depend on reflexes alone, but involves learned voluntary movement as well. When we realize that movement experiences shape our body, that should be our clue to connect riding a bike, walking, jumping and skipping with bending at the sink to shave, getting out of a car or carrying out the trash: they are reflex and voluntary. Postural response is shaped by experience.

Voluntary movement is movement with a purpose.[2]

Daily movements are important since they require skill much as a great tennis serve or the freestyle performed by an Olympic swimmer requires skill.

The trick is to get movement to stick so that it's a "no brainer." We do that by "writing" a motor engram, a template that takes us from frustration to freedom by saving a move in our long-term memory.[3]

We "write" the engram by practice, by repetition. Consider this. It takes 5,000 to 6,000 repetitions to retrain the muscles of our body to coordinate as they did when we were a child.[4] Before you put down this book and give up, realize that in one day you probably bend forward hundreds of times. In one day!

This is what repetition brings:[5]
- To net a basketball from any angle takes 1 million throws to reach peak skill.
- It takes 1.5 million stitches to be a fine hand knitter.
- A baseball pitcher needs 1.6 million throws to be a Nolan Ryan.

So what's a few thousand bends? It's all in your perspective. However, if you're going to take the time to do your thousand repetitions, make sure you are tuned in to how your muscles work.

Muscle mania

The lean long legs with the big broad back is how motion creates commotion in this day and time of "being fit." This is all very well but if those fleshy parts do not know how to move well or naturally, it doesn't matter how well you can prance across the room; their health is in the how.

Since muscles make joints act, their main roles are to maintain posture and generate force. Skeletal muscle fibers work because of their connection to a specific motor neuron, which makes a motor unit, which makes the muscle move.[6]

When a motor neuron fires, all of the muscle fibers innervated by that neuron will contract at the same time. The quadriceps muscles (thigh muscles) have large motor units which means they have several thousand fibers connected to one motor neuron, whereas the muscles of the hand may have only 10 fibers with one motor neuron producing more fine-movement patterns.[7]

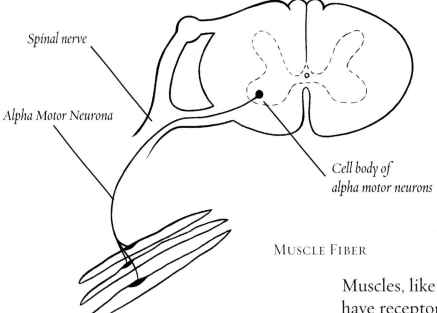

Spinal nerve

Alpha Motor Neurona

Cell body of alpha motor neurons

MUSCLE FIBER

Muscles, like joints, have receptors that pick up and relay information on how the body is moving.[9] Research has shown there is a high density of these receptors in the postural muscles of the spine.[10]

Job Descriptions

Muscles are labeled by their location and job description.

Tonic or postural muscles come into play with low force activities like daily movements. They maintain joint integrity by controlling movement at the joint site. As the force of contraction increases, other muscles are progressively recruited. These muscles are called phasics and continue the movement that the tonics begin. The tonics are the "stayers" and help hold the body upright and in place. The phasics are the "sprinters" which help move parts of the body rapidly for short periods of time but will fatigue quickly. Your face and eyes have more phasic muscles that do quick movements and then rest for a time. The phasic muscles are not expected to continue to contract over a long period of time without rest.[8]

Your face and eyes have more phasic muscles that do quick movements and then rest for a time.

Showgirls

To understand the healthy interrelationship of the tonic and phasic muscle groups think about showgirls and their high-kick routine. Harmony between these two types of muscles keeps their legs high in the air without their bodies falling backwards. The tonic (postural) muscles keep the girls upright while the phasics enable beautiful long legs to kick up high.

Although we may not perform high kicks daily, we do reach with our arms to get a cup off a high shelf, and we do walk, bend and turn: all movements that need tonic/phasic interplay.

Muscles, like joints, have receptors that pick up and relay information on how the body is moving.[9] Research has shown there is a high density of these receptors in the postural muscles of the spine.[10]

The tonic (postural) muscles keep the girls upright while the phasic muscles enable beautiful long legs to kick high in the air.

Why doesn't our body tell us we are messing up? Once we move off of our natural axis of motion, we lose the ability to know how to move correctly because the receptors cannot pick up the signals.

Nina . . . can make her body undulate through her incredible control of muscles most of us ignore (can you roll quarters down your belly?)

These receptors enable your body to tune in to the spinal muscles with great sensitivity. This level of detail, plus coordination of muscle groups, lets the show-girls know how to stay stable as they do high kicks. It's also how Nina, the belly dancer, can perform what seems a totally opposite type of move. She can make her body undulate through her incredible control of muscles most of us ignore (can you roll quarters down your belly?) because the muscles deep in her back know how to work with the larger mus-cle groups of her torso. This relaxed but powerful muscle control begins deep within.

It is important for these two muscle groups to work in harmony, especially in the lower trunk and spine area, which is the power center where movement begins.

If the tonic muscles cannot do their job, then joints break down. This does not stop us from moving because we substitute other muscles to help us do a desired task. However, using substitutes day in and day out slowly wears and tears the joints and tissue, creating weakness in tonic muscles because of breakdown between the two systems.

This relaxed but powerful muscle control begins deep within.

Jody has her pelvis forward and her arms across her chest. Her head is forward, one knee is locked and the other foot is turned out in front. She is "hanging" on her ligaments.

Why doesn't our body tell us we are messing up? Once we move off of our natural axis of motion, we lose the ability to know how to move correctly because the receptors cannot pick up the signals. The body is very forgiving and we can cheat, without knowing it, until damage is already done.

How We Cheat with Our Muscles

A good example of how moving off the axis causes us to "cheat" and use other muscles which eventually leads to damage is the typical posture of most female adults.

Jody has her pelvis forward and her arms across her chest. Her head is forward, one knee is locked and the other foot is turned out in front. She is "hanging" on her ligaments. Why? Because many of her postural or tonic muscles are dormant–she "hangs" instead of stands. Compare this with my relaxed but upright stance on the next page. Jody will also walk and move differently from me because her receptors cannot pick up the information to make the right choices in movement. This will result in fatigue of her postural muscles–no matter how good a personal trainer she has keeping her "fit."

strength

Strength can be defined as either the size of the cross-sectional area of a muscle (big biceps) or the ability to do more work.[11]

Research confirms my theory on strength, defined functionally: "If the patient can do more work, i.e., lift a heavier weight, yet does not have hypertrophied muscle fibers [bigger muscles], then this implies an increase in the efficiency of the patient's neurological system. This is called neurological adaptation and is a function of coordination."[12] That is a very scientific way to explain why I am skinny but STRONG!

Neurological adaptation is gained from the *Body Balance Method*. It is what is greatly missing in our culture–strength achieved by training our neurological systems to do the specific task at hand, i.e., daily movements.

Myth: Understanding Your Abdominal muscles

The mighty abdominals are the most misunderstood muscles in the body.

Let's kill the greatest myth with one blow. How does holding in your stomach by contracting your abdominals and doing stomach crunches protect your back?

If you have any knowledge of how the body works you will know that back muscles are, in general, responsible for your back (now that was profound).

Lisa Ann's stance, however, is upright but relaxed.

Doing any type of stomach crunches is similar to slumping in a chair. Bringing the body OUT of balance.

There are also many strong ligaments, structures and connective tissue responsible for protecting the back. Given a chance, the abdominals working with other trunk muscles do a great job of holding up your back! If by some freak chance the myth (holding in your abs) was true, then people with bad posture would have great backs because they are constantly in the abdominal crunch position. Abdominal muscles do not work in isolation but in conjunction with the back muscles, both statically and dynamically. If you do abdominal crunches, your abdominals adapt to a more shortened position creating functional weakness and changing the position of the pelvis. When your abdominals are relaxed and allowed to go back to their natural length, then any movement involving the trunk can be your "ab" workout.

If you are to get oxygen to the muscles, you need your diaphragm to be able to drop as you inhale. If you hold our stomach in, you are contracting your abdominals, which restricts you from taking a normal deep breath. For comparison, take the muscles of the wrist and hand. If you held your hand in a tight fist all day, wouldn't that seem odd and damaging to the joints and muscles of your hand? So why does one of the strongest muscle groups in your body need "holding" to be healthy when it obviously is not logical?

I may have convinced you logically, but I know that many of you are thinking: "Lady, if I let my gut hang out, it will look so bad."

The interesting thing is, as thin as I am, my gut looked really bad when I first began to let it go. It looked like a bowling ball directly out in front of me. It may not look good, initially, when you begin to relax your abdominal muscles. First, the muscle is weak and does not have its natural tone because it has not been used correctly. It is also short because the skeletal structure is still bent forward. It is even more pronounced due to the position of the spine. Third, it is not healthy tissue because it has not received the correct amount of stimulus to move as it could and should.

It is time to stimulate collagen as it was designed to be stimulated and watch your abdominal area tone up with the rest of your body. One of the first sensations of standing correctly is feeling the lengthening of the spine as the pelvis drops, and creating a firmness of the abdominals as they lengthen. If it is fat that you are hiding by holding in your abs, well, *Body Balance* is the best way to deal with it and start moving more. The most common response my patients have to their new posture is that other people think they have lost weight when they have not lost an ounce. So, begin to rearrange your tissue and see what people say!

Strength training: increase the resistance in a natural way

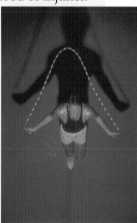

If we begin to focus on strength training in a more natural way, the overall results will be more beneficial. When the multiple systems of the body work together as designed there is less likelihood of injuries. Rather than concentrating on getting stronger, think of gaining strength as part of something you enjoy. You could swim, rock climb, fence, row, do racquet sports, dance, play basketball, volleyball, box, jump rope, hike or kayak. The list goes on and on. Your mind and body would be so much more excited by a variety of activities. You'll not only gain strength, but also flexibility, coordination and mental satisfaction. Your body is not designed to work in a segregated fashion.

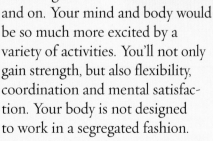

stretching

People often call me and say they need a "stretching program" because they feel "tight" and can't touch their toes. (That, by the way, is what your knees are for: if you want to touch your toes, bend your knees! It is a lot easier that way.)

When a muscle works, it contracts, creating a force. This force can be influenced by a number of factors, one being the length of the muscle. For all muscle there is an optimal length for force generation. With *Body Balance*, optimal positioning of the muscles and their joints is key to creating optimal length in muscle.

So, what does stretching tight muscles really mean?

Most people do not realize that some parts of the skeletal muscles contract and some parts do not. Muscle is made up of two types of tissue: contractile and noncontractile. Contractile tissue, when activated, generates a force or tension.[13] Noncontractile tissue is the connective tissue in muscle that transmits the contraction forces to adjacent structures, such as bones or tendons.[14] It also provides routes for nerves and blood vessels.[15]

Instead of thinking a muscle needs stretching, think that it needs to lengthen in a way that it is to be used. If I repeatedly bend deeply from the hips (as in the picture) my muscles adapt to this length. If I kept bending at the waist instead, and sat on my low back and tailbone, my hamstrings would adapt to move through a shorter range. This shortening would make them feel

Butt Up the Wall

. . . if you want to touch your toes, bend your knees! It is a lot easier that way.

tight if I then tried to move past the range they were used to moving.

A muscle adapts to a certain length depending on how it is used. If you feel like you are stretching your hamstrings when you bend, you are actually contracting your hamstrings even though the sensation feels like they are stretching.

"Stretch" is something we feel when we move our body through unaccustomed ranges.

People often mistake my work for a stretching routine. I have heard people say "she really stretches you." I do not think people realize that when the muscles and joints move as designed a person does not feel tight. Interpret the word "stretch" as a feeling that occurs through the process of the joints and bones moving back to their optimal position.

For example, when learning to do the most popular move, Butt Up the Wall, (opposite page) people think it is a great hamstring stretch. You will feel a sensation of stretch in your hamstrings when you do that exercise, but it is really an eccentric contraction* of your hamstrings. What makes your hamstrings "stretch" or lengthen is using Butt Up the Wall repeatedly. This means that your hamstrings are adapting to a more lengthened position because of the way they are being used. As you integrate this movement into your daily life (it is how to bend naturally, which we do hundreds of times a day), your back, buttocks and neck become stronger. All this results from just Butt Up the Wall.

Stretching a Calf Muscle Could Kill Your Jump

Connective tissue stores energy just like a vaulting pole. If you aggressively bend that pole, it has a reactive force that propels you through the air. It's the same with your Achilles tendon. When you prepare to jump, your calf muscle shortens and your tendon extends. The extension process absorbs energy, just like the vaulting pole, and springs you into the air while your calf muscle stays in a tightly shortened position to create the most force. If you stretch a basketball player's calf muscle, his tendon will lose some of its spring and his calf muscle will generate much less force, because it has increased in length as well. This could kill his jump.[16]

Your Structure

Bone is alive. It is a constantly changing, self-repairing tissue. Through a remodeling process, under normal conditions bone resorbs and is replaced by newly formed bone.[17]

There are different bone structures. There's a beautiful lattice-like structure (trabecular bone) with a three-dimensional design. Its spacious surface area helps it resist compression forces. This bone type is found in the vertebral body and in the end of long bones. The smooth strong walls of the cortical or compact bone are well placed in the midsection long bones of the extremities. They respond to bending and swaying.[18]

Bone stays healthy and alive with movement and good stress. Too much load, or load in the wrong direction, can disturb the structural foundation of bone. *Body Balance* moves keep your posture and alignment in place to prevent structural changes.

The quality and quantity of the forces on bone are important. They need to be directly related to the way the bones are stacked in relation to gravity, as well as how the bones move via the muscles. This concept is based on Wolff's law: "Bone will change its internal architecture according to the forces placed upon it."[19]

What happens if not enough quality weight is placed on bone? It internally weakens, making it more susceptible to injury. The mineral content will decrease and the bone becomes weak without enough weight-bearing activity. This common problem, known as osteoporosis, is seen most often occurring in the lattice-type bone of the vertebral bodies and the head of the femur. These are the major weight-bearing bones of the skeletal structure. While various factors play a role in

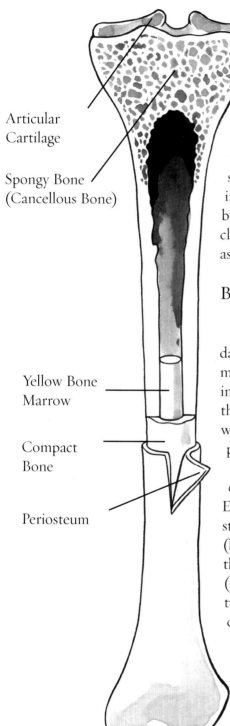

Articular
Cartilage

Spongy Bone
(Cancellous Bone)

Yellow Bone
Marrow

Compact
Bone

Periosteum

this breakdown of bone tissue, one area that *Body Balance* can address is increasing the amount of CORRECT stress on the bone. Correct stress means there has to be more than just generic force on bone. Generic force is weight-bearing exercises without the skill of the movement integrated. Of primary importance is vertical weight bearing through bones though the mechanical stimulus of muscle contracting against bone surfaces is important as well.

Build Bone As You Walk

Women who exercise by walking two miles a day and have good alignment with gravity will get more quality compression on their weight-bearing bones and better mechanical stimulus from their back muscles. Women who do not move well don't get maximum benefit because of their postural misalignment.

A study in 1982 by Bloom and Pogrund on osteoporosis compared African Bantu and English women in different environments.[20] This study found that both groups suffered arm (humerus) bone loss at a similar rate. However, the African women had a low rate of hip (femoral neck) fractures. In the U.S., hip fractures are common due to osteoporosis. The key difference seems to be that the African women remained more active, bending and stooping more, which preserved their hip bones.

Contrast these two Portuguese men. In the background this man's posture represents the older generation, with hips under his spine creating a stable support to move from.

The man in the foreground has the "new generation" posture with his hips tucked under and in front of his spine, vulnerable to injury.

"In all things, success depends upon previous preparation without which there is sure to be failure."
Confucius, 551-479 B.C.

The researchers said: "One further factor which has not to our knowledge been previously discussed is the influence of posture and mode of walking. The female Bantu walks with stately posture carefully placing one foot in front of the other, often balancing large parcels on her head, and rarely hurries. It is our feeling that this factor is of some importance in reducing the rate of femoral neck fracture, as minimal trauma due to falls while walking is often the cause of such an accident in more developed societies."[21]

This slow mode of walking is not easy; a correct slow walk is a balancing act. Walking and bending as I advocate (and as the Bantu women do) can be a preventive measure as well as a treatment for osteoporosis.

There is another system that helps us stay upright: the vestibular system in our inner ear. This system is made up of hair cells bathed in fluid throughout the inner ear. It works with the sensory systems of our body. We could not walk and keep our balance without the vestibular system.

This system also gives us the ability to stabilize our vision while we move so that objects stay still. Otherwise, every time we move, our world would move too.

One of the two main systems of the vestibular system is responsible for coordinating movements or movement strategies (the Vestibular Spinal Reflex). The *Body Balance* moves get your body's receptors working efficiently, which improves the interaction of your body with the vestibular system.[22]

Conclusion

Chapters 6, 7 and 8 show how much our bodies and minds are integrated. They illustrate that all of the parts, from the smallest chemical released from your hypothalamus to the thousands of muscle fibers contracting as you run, are part of the whole experience of daily movement. This understanding will help you relate to how the movements in Chapter 4 rebuild your relationship to every movement in your life. You will never simply sit down mindlessly again. You will know how every move affects all of you. With this insight, you can reclaim the secret of a lost art.

A belly laugh a day keeps the doctor away. Give the phasic muscles of your face a good workout by LAUGHING.

1 E. Kandel, J. H. Schwartz, and T. M. Jessell, *Principles of Neural Science*, Appleton & Lange, Connecticut, 1991, p. 534.

2 M. Trew and T. Everett (eds.), *Human Movement*, Churchill Livingstone Inc., New York, 1997, p. 85.

3 Ibid.

4 O. Holten, H. P. Faugli, *Medisinsk Treningsterapi*, Illniversitetsforlaaet. 0608 Oslo, Norway.

5 F. J. Kottke, D. Halpern, J. K. M. Easton, A. T. Ozel, C. A. Burrill, The training coordination. Archives of *Physical Medicine and Rehabilitation*, 1978, 59: 567-572. M. Trew and T. Everett (eds.), *Human Movement*, Churchill Livingstone Inc., New York, 1997, p. 93, cited by second source given.

6 M. Trew and T. Everett (eds.), *Human Movement*, Churchill Livingstone Inc., New York, 1997, p. 55.

7 Ibid., pp. 54-55.

8 C. Z. Hinkle, *Fundamentals of Anatomy of Movement*, Mosby, Missouri, 1997, pp. 82, 83.

9 A. C. Guyton, *Textbook of Medical Physiology*, W. B. Saunders Company, Pennsylvania, 1986, pp. 6-8.

10 Nitz et al.; B. Brownstein and S. Bronner (eds.), *Functional Movement in Orthopaedic & Sports Physical Therapy: Evaluation, Treatment & Outcomes*, ed., Churchill Livingstone Inc., New York, 1997, p. 43.

11 D. Lamb, *Physiology of Exercise*, MacMillan, p. 260.

12 Ola Grimsby, *Clinical & Scientific Rationale for Modern Manual Therapy*, MT-1, Independent Study Manual, San Diego, 1998, *Exercise Physiology*, p. 17.

13 M. Trew and T. Everett (eds.), *Human Movement*, Churchill Livingstone Inc., New York, 1997, p. 46.

14 Ola Grimsby, *Clinical & Scientific Rationale for Modern Manual Therapy*, MT-1, Independent Study Manual, San Diego, 1998, *Exercise Physiology*, p. 12.

15 Ibid.

16 Boyson-Moller; Ola Grimsby, *Clinical & Scientific Rationale for Modern Manual Therapy*, MT-1, Independent Study Manual, San Diego, 1998, p. 36.

17 D. Currier and R. Nelson, *Dynamics of Human Biologic Tissues*, F.A. Davis Company, Philadelphia, 1992, p. 258.

18 J. A. Gould and G. J. Davies (eds.), *Orthopaedic and Sports*, C.V. Mosby Company, Missouri, 1985, pp. 28, 29.

19 Wolf, J., *Gestezkter Transformation der Knocken*, Berlin, A. Hirschwald, 1884.

20 R. A. Bloom and H. Pogrund, "Humeral Cortical Thickness in Female Bantu "Its Relationship to the Incidence of Femoral Neck Fracture," *Skeletal Radiology* (1982), 8:59-62.

21 Ibid., p. 62.

22 Bridgette Wallace P.T., interviewed by Lisa Ann McCall, June 2000, Dallas, Texas, phone conversation.

* There are three types of muscle contraction. When you bend correctly you do an eccentric (lengthening) contraction of your hamstrings; if you stay there for a moment it's an isometric (no change in length) contraction and when you come up it is a concentric (shortening) contraction. So by bending correctly you have worked your hamstrings three different ways.

CHAPTER IX
RESHAPING YOUR WORLD

CHAPTER IX

YOUR ROLE IN RESHAPING YOUR WORLD

Now you know about Body Balance, *but where do you start to incorporate it in your life? This chapter tells you how to start right now and encourages you to take personal responsibility for your body. This isn't just good for your health but for your pocketbook.*

Reshape Your World

What's the "bottom line" to a good-looking body? If you want to feel good and look good, think of the old saying: "know thyself." How you view YOU is the key to good looks. Begin by having a healthy view of what constitutes a good-looking body. Next, be realistic about the changes you can make and make the change process enjoyable. It starts with feeling good about yourself. Let your body sculpt itself into its shape with your focus on getting the shape it was built to have. It's a self discovery. Examine your day-to-day life right now. As you sit on the couch, you may think the health club holds the answer to getting into shape. Or perhaps you believe that without the correct gear you can't peel off the pounds.

"If I could only get my inner inner thigh a little smaller"

The Hungry Life

The real answer lies in how you get off the couch and how you perform those everyday chores. These mindless movements, done correctly, are the secret doorway to health.

As you move across the room, you can bend in a fashion that builds a better back and buttocks. Let the stairs in your house become "step class"—without someone yelling at you to do more. Get something done while working up a sweat. Take out the trash, or unload the dishwasher. You have a built-in gym right at home. Learning how to get more out of each stride while walking Spot can firm you up as you discover the secret to moving better: a brain/body workout. Carry the groceries, take the stairs at the office and look for reasons to move more.

The Full Life

Look around and you'll see that we've lost the art of moving properly. It shows up not only in the way we move, but the way we've grown as a nation.

Obesity is a national crisis in America, so don't think you are alone in the fat war. More than half the population is overweight, and one in four of us is obese, which is defined as carrying an

You do not have to be a statistic IF you use what you already have.

extra 30 pounds.[1] Just 100 unused calories a day can add up to more than 10 pounds a year.[2]

The cost of carrying these extra pounds is estimated to be some $238 billion annually. This just covers the health care costs of obesity, but does not take losses of employee productivity into account.[3] You do not have to be a statistic IF you use what you already have.

Q & A

Dr Condry: "I see a lot of people with different aches and pains, fatigue and stress from their lives and jobs. For a lot of these people, I think that bad nutrition and really poor lifestyles contribute in a major way."

Dallas gynecologist Dr. Lessa Condry believes obesity is closely connected to today's lifestyle and how women care for their bodies. This is how she put it when I interviewed her recently.

L A McCall: "If you could change three things in the lifestyles of women what would they be?"

Dr Condry: "It would be nutrition, exercise and some sort of mental hygiene, whether it's religion, meditation or some other spiritual activity that helps their lives."

LAM: "If we could put a dent in the price tag of medicine, what do you think would be the best attack or most effective change?"

Dr Condry: "Exercise, exercise, exercise."

LAM: "What exercise would you suggest?"

Dr Condry: "Because obesity is such a problem, I just want them to do something. Just move as opposed to doing absolutely nothing, which is what they do right now. Patients always complain about their weight, but if they want to change how they look, they will have to do more aerobic types of exercise. We need to start somewhere."

LAM: "Do you think the way we move could make changes in orthopedic-related problems such as neck, back or shoulder ailments?"

Dr Condry: "Yes. Probably the biggest problems I see in women are neck, upper back and sciatic pain and these are probably related to posture during work."

LAM: "What is the percentage of women you see with neck, back or soft tissue problems?"

Dr Condry: "I'd say 50% complain of some problem like that in just the course of an annual exam."

LAM: "Sum up your thoughts on the importance of movement or the quality of movement plus its role in women's health and the longevity of their lives."

Dr Condry: "In terms of osteoporosis, certainly your posture makes a huge difference. Exercise and movement is very important in maintaining the strength of bone. In my patients, those 70 to 80 years old, the ones who have exercised and kept moving are the healthiest ones.

American women are plagued by problems of genital prolapse as they get older. I wonder how much our posture and the way we tuck in our stomachs may have to do with increasing abdominal pressure and causing the pelvic organs to fall out. The uterus distends and can ultimately come out of the vagina. Some of these are childbirth injuries, but they are made worse because women don't do enough strengthening exercises as they age."

A Little
Goes
a Long
Way

A Little Goes a Long Way

Less than 20 percent of U.S. adults are involved in a vigorous, sustained exercise routine on a regular basis. This is nothing new—we were just as inactive in the '80s. This sedentary lifestyle not only affects our size but can be directly related to heart disease, Type 2 diabetes and colon cancer.[4]

If you're not motivated to work out in a gym, maybe the old-fashioned way of moving to stay in shape could be your answer. Research shows that a regular brisk walk may be as good as a structured workout at the club.

Two groups were studied over a 24-month period. Group 1 followed a traditional structured exercise program while Group 2 participated in a "lifestyle" physical activity program, which means the individuals increased their physical activity to a moderate intensity as part of their daily routine.[5] The structured exercise people spent 20-60 minutes 3 to 5 days a week exercising with a moderate-to-high intensity workout program. The "lifestyle" people spent at least 30 minutes in moderate-intensity physical activity, preferably each day, in a way that fit into their unique lifestyles.[6]

The results showed that they both produced "significant and comparable changes."[7] These changes included improvements in cardiorespiratory fitness, blood pressure, body fat percentage and physical activity.[8]

It may be more realistic to increase your activity to a moderate degree for 30 minutes almost every day and stay with it over a *lifetime*. That means you become involved with movement on a moment-by-moment basis.

Research shows that a regular brisk walk may be as good as a structured workout at the club.

Q & A

DR JOHNSTON: "The increase in back problems I put down to lack of activity, being overweight and sitting at computers all day. As far as shoulders and elbows are concerned, these injuries are more related to activity. More people are playing sports that are stressful to elbows and shoulders. I see fewer knee problems than I did 20 years ago when everybody was running."

LAM: "You are relating the increase to various sport activities?"

DR JOHNSTON: "Yes, in this practice, because of the more active population."

LAM: "Do you think that knowing how to move better helps these problems?"

DR JOHNSTON: "Absolutely. People need to learn the right way to be active and not rely on outdated notions."

LAM: "What do you consider is the way to control problems with weight, stress, blood pressure, cholesterol?"

DR JOHNSTON: "One of the problems is that people have a perception that exercise means doing a structured program 3 or 4 times a week for 30 minutes of aerobics. While that's good, most people can't attain

One internist who advocates this exercise regime is Dr. Richard Johnston. His busy practice is heavily weighted toward corporate types, 35 to 50 years old. He has seen a noticeable rise in patients with joint pain and dysfunction during the past decade. The biggest orthopedic problem he sees is back related, followed by shoulder, knee, and elbow ailments. I asked his opinion for this increase.

that. If they would just move, they'd be better off. They don't necessarily have to go exercise every day, they just need to move around, walk up stairs, or park farther away. Fortunately, the pendulum is swinging back and we are getting back to walking.

LAM: "How is posture important in our health now and in our future?"

DR JOHNSTON: "On any given day, half of what I see could be prevented by just taking care of yourself. And short of traumatic events, all of what I see—be it soft tissue or mechanical problems— could be taken care of through correct movement, posture and a reasonable amount of exercise. "In their general checkup, about half my patients will complain of some joint problem. Remember, we're talking about 35 to 50 year olds. If I had an older practice, it would be 100 percent. Most of these complaints could be handled by the individual. The savings would be astronomical on backs alone!

"Most patients with musculoskeletal problems will probably get an x-ray whether there was trauma or not, and will usually be prescribed a nonsteroidal anti-inflammatory costing anywhere from $60 to $100. Of these, 20 percent will get side effects from drugs. Patients may be sent to a physical therapist and be started on an exercise program that they could have done on their own. Finally, add an MRI costing $1,500."

LAM: "Estimate the amount of drugs you prescribe just for joint dysfunction pain in a week."

DR JOHNSTON: "If you're talking about anti-inflammatories and muscle relaxants, the total cost would be several hundred dollars a week and I'm on the conservative end. I know doctors who immediately start physical therapy and get an MRI. There you've just added several thousand dollars to one individual for no change in outcome."

Reshape
Your Pocketbook

Four out of five American adults will experience back pain. Two aspects can hurt: the physical pain and the financial pain.

Insurance does not always cover everything, but using what you already have is an insurance plan and a preventive measure. And it's cheap. Moving in comfort is more than a luxury; it could prevent that hip replacement or rotator cuff tear or even a nagging backache.

You'll recall that backaches alone cost this country an estimated $50 billion annually.[9] Osteoporosis costs us a further $10 billion a year nation-wide. Let's come down to an individual basis and look at some prices of having things fixed in the year 2000.

A total knee replacement costs $3,000 to $4,000;
A total hip replacement costs $2,000 to $3,000;
A major knee reconstruction is $2,000 to $7,000;
Arthroscopy of knee $1,000 to $3,000;
A discectomy of the back $2,500;
A fusion of multiple levels of the spine $4,000 to $12,000;
Scoliosis surgery $14,000 to $17,500 (not counting a five day hospital stay);
Arthroscopy of the shoulder costs $2,500 to $3,500;
Rotator cuff repair surgery is $2,500 to $5,000;
An emergency room visit for a hip dislocation is $7,000;
The hospital bill for a hip revision (five days) costs over $27,000.

These prices do not include the operating room cost and the anesthesiologist fees.
(The price tag may vary depending on your physician and your location.)

A constant cash Flow: Moving Skillfully

Think of relearning to move daily as money in the bank.

If we moved regularly, the way we were designed, we would save so much of our energy fund that we could have energy left over. As it is now, we are physically in debt before the age of 40.

Remember what we learned from the Bantu study (see Chapter 2). By their mid-30s, Europeans had only 10 degrees of forward-and-backward motion in their lower back spine while Bantus had kept a "younger" spine because they had retained twice as much motion in this area. This means that by the time you are 40, your back is much "older" than it needs to be.

When moving without awareness we invest our "energy funds" in ways that cost more and we get less. Your golf, running or tennis game may end because of the debt your body has acquired. Loss of skill may cost you only a few dollars each time but it adds up over the years. We drain our physical account by not using what we already have.

Q & A

DR JONES: "My specialty is joint restoration—notice, it's not total joint replacement."

LAM: "How many hip replacements do you perform in a year?"

DR JONES: "500 hip and knee replacements in a year and that's a huge number."

LAM: "What is the main cause of a need of surgery?"

DR JONES: "Males and females are different. Generally what we're dealing with is degenerative arthritis, which is the wear and tear on the bone, and part of the wear can be high impact loading. Someone who is overweight is loading the joints more than someone who is not overweight."

LAM: "How long do hip replacements last?"

DR JONES: "Statistically, the middle of the bell curve is 10 to 12 years, but there are many individual varying factors. The biggest negative factor for males is being less than 60 and being overweight. Males, because they just pound their hips more; and less than 60, because they think they're bulletproof and that they still have their own young joint. And the weight, for obvious reasons, puts more stress on them."

Dr. Dicky Jones believes personal responsibility is the sensible way to regain—and maintain—good health. He's in the repair business, replacing hips and knees on a daily basis. The aim at his clinic, he says, is to help people understand that they are responsible for the choices they make, and what health professionals do is to guide them through their choices. Dr. Jones explains it in his own words.

LAM: "Have there been any age-related changes in the past 10 to 15 years for men and women needing hip replacement? Like a younger age group coming in?"

DR JONES: "If you talk about just hips no. We're seeing knees because people are not willing to put up with a knee that doesn't function very well. I've got guys that come in here, and if they can't walk nine holes, they'll want a new knee. Now, do I operate on them? No. I tell them to get on a bike and start riding. The vast majority of them do quite well. They can learn to live with their own joints because you're always better with your own joints."

LAM: "What about knee replacements for women?"

DR JONES: "The number one reason is obesity. The body mass index is directly related to osteoarthritis of the knee."

LAM: "Does movement play a role in the longevity of hip or knee joints?"

DR JONES: "Without any question the more mobile and flexible a joint can be, the longer it will last. Once arthritis begins, secondary effects occur, such as inflammation, therefore irritation, and then pain. When that occurs, we are dealing with a joint that doesn't want to move fully. Then the stresses become concentrated and the more the stress is concentrated, the more there's a problem. This causes more inflammation and more wear and tear. If you can maintain full range in your knee, no matter how much previous trauma you've had, generally you can maintain your own knee. You can lessen the impact by wearing the right shoes and by controlling your weight. You can also enhance the nutrition of the cartilage by maintaining good flexibility of the joint with some sort of exercise, like biking.

"The way that we work here is to help people understand that they are responsible for themselves. I like what you're doing because I think you are doing the same—helping them understand that what they do, the choices they make, and the ways that they do things can bring very positive consequences and help them enjoy life. That's not saying that they're necessarily going to have a longer life, but their quality of life as they age will be considerably better."

LAM: "It is obvious that we have a problem with cost in medicine since the price tag is way too high. What do you think would make the biggest dent in the price tag of our medical problems today?"

DR JONES: "More people like you helping people to understand that they can be responsible for their own healthy choices and their good behavior. And more people like me doing the same thing."

In 1995 there were 134,000 hip replacements and 216,000 knee replacement surgeries in the United States.[10] If we could make a small dent in these numbers it would begin to liberate us from costly insurance premiums and help us regain control of the madness in medical care today. Just think about what it would do for our mental health when we realize we hold the answers to so many of our problems in our hands.

It is said that the two things individuals concern themselves with as they get older are health and money. Sometimes it's the good health going bad that drains the bank account.

Q & A

DR MONTGOMERY: "Crazy sports! Here in Texas, football is a collision sport and we see a lot of injuries. Another is snow skiing, probably the most dangerous sport in the United States for time spent. The rest of them are women's sports. We see a large proportion of problems in women's sports such as basketball, some volleyball and cheerleading. I probably see two women for every man."

LAM: "Why do you think women are needing more knee surgery than men?"

DR MONTGOMERY: "I don't think there's any doubt but that it's all secondary to core mechanics: pelvic mechanics, lumbar spine mechanics and the inability of the muscles to work off a good base. Therefore the quadriceps and the hamstrings cannot protect the ligaments in sport."

LAM: "Do you think hormones play a role?"

DR MONTGOMERY: "I think there's some evidence that hormones play a role but I really think instability of the pelvis and lumbar spine is the biggest culprit."

LAM: "Do you think the growth factors, such as growth spurts around age 15 make a difference?"

DR MONTGOMERY: "It's interesting that 10- or 11-year-old females are probably much better athletes than 14 year olds. As they go through a growth spurt, it changes their center of balance. And because of that, they go through a stage of poor coordination where they just don't have good muscle control."

LAM: "How many people come through your office whom you can't fix with surgery?"

DR MONTGOMERY: "I'll see approximately 40 new patients a week, and I'll operate on 8 to 10 of those patients. So 75 percent of the problems I see are non-operable."

LAM: "Could those people whom you can't help with operations benefit by changing their daily postures and movement?"

DR MONTGOMERY: "I think it's all about balance, posture and mechanics. Everyone, if they really worked at it, would improve probably at least 50 percent by becoming aware of their movements and their balance. The problem we see is that people want to take a pill or get a quick fix. Of all the people whom I see, probably 25 percent will really benefit but that's only because I feel most people won't stick to a program that rehabil-

Dr. Jim Montgomery has always been involved with young people and sports. This leading orthopedic surgeon was head physician for the U.S. Olympic Team in Barcelona in 1992, is chairman of Medical Services for the 2000 U.S. Olympic Team and is clinical associate professor of orthopedics at the University of Texas Southwestern Medical School. He has specialized in knees for a dozen years, performing more than 550 knee surgeries annually. I asked him for the main cause of these surgeries.

itates and decreases the amount of problems they have."

LAM: "Give me your thoughts on knees in general from your experience in the Olympics, working with professional athletes and then working with everyday people. Is there a universal principle?"

DR MONTGOMERY: "I think that we in America have missed what the Japanese and Chinese said over 2,000 years ago and have practiced since that time: we need to balance our minds and balance our bodies."

Most Important Natural Resource: Young People

Our young people are our most important natural resource for the future. Why should they have to pay unnecessary prices daily that will affect their health down the road? As mentioned in Chapter 5, the *Body Balance* principles can change the future of medical costs for everyone if we address this reshaping of our bodies.

A Paradigm Shift

A Paradigm Shift

To correct structural and functional problems that plague the western world will take an entire shift in perspective and approach. But it is doable—look what we've done with our approach to heart disease. In recent years we have seen how we can reverse problems such as heart disease, blood pressure and cholesterol through diet, exercise and stress management. This type of approach works when people change from being a spectator being treated by an outside source to being the primary care "doctor" prescribing quality lifestyle changes for themselves. This means we all must look inside ourselves to get the desired outcome.

The McCall Body Balance Method asks you to stop being a robot as far as your everyday moves are concerned. Use every move you make to regain health and keep you fit. *Body Balance's* integrated approach puts life back into whatever you do. It shows you how to stop tissue breakdown, rid yourself of pain, build better bone and improve your balance all by returning to the basics of how you were designed to move.

Being comfortable in your body and enjoying it moment by moment is the magic we're missing in daily life. It is more of a transformation than a treatment because the key ingredient is to have fun while you're doing it. Life was meant to be lived, not endured!

Today

Today take the long road to everything. Burn more calories without changing your clothes or schedule. Take the stairs instead of elevator. Walk to lunch in-stead of riding or driving. Do a few chores around the house. It's like money in the bank!

1 L. Beil, "Girth of a Nation: Doctors warn of health crisis as obesity gains on Americans," *The Dallas Morning News*, August 29, 1999, p. 1.

2 Ibid., p. 2.

3 E. Gleick, "Land of the Fat," *Time International*, October 25, 1999, p. 2.

4 Dunn, B. H. Marcus, J. B. Kampert, M. E. Garcia, H. W. Kohl, and S. N. Blair, "Comparison of Lifestyle and Structured Interventions to Increase Physical Activity and Cardiorespiratory Fitness," *The Journal of American Medical Association*, January 27, 1999, Vol. 281, p. 327.

5 Ibid.

6 Ibid., p. 329.

7 Ibid., p. 332.

8 Ibid., pp. 332-333.

9 Richard Deyo, "Low-Back Pain," *Scientific American*, August 1998, p. 98.

10 *American Academy of Orthopaedic Surgeons Facts about Total Hip and Total Knee Replacement*, p. 1.

Bibliography

Armstrong, T., *Seven Kinds of Smart*, Penguin, 1993, p. 82.

Beil, L., "Girth of a Nation: Doctors warn of health crisis as obesity gains on Americans," *The Dallas Morning News*, August 29, 1999, p. 1.

Benson, H., *Beyond the Relaxation Response*, Time Books, New York, 1984.

Bloom, R. A., Pogrund H., "Humeral Cortical Thickness in Female Bantu–Its Relationship to the Incidence of Femoral Neck Fracture," *Skeletal Radiology* (1982), 8:59-62, pp. 2-5.

Bogduk, N., Twomey L., *Clinical Anatomy of the Lumbar Spine*, Churchill Livingstone, Melbourne, 1991, pp. 7, 66.

Brand, R., "Hypothesis-Based Research," *The Journal of Orthopaedic & Sports Physical Therapy*, August 1998, Vol. 28, No. 2, pp. 71/72.

Brownstein, B., Bronner S. (eds.), *Functional Movement in Orthopaedic & Sports Physical Therapy: Evaluation, Treatment & Outcomes*, ed., Churchill Livingstone Inc.,New York, 1997, p. 231.

Cech, D., Martin S., *Functional Movement Development Across the Life Span*, W.B. Saunders Company, Pennsylvania, 1995.

Colt, G. H., "The Magic of Touch," *Life* magazine, August, 1997, pp. 7, 15, 60, 62.

Csikszentmihalyi, M., *Flow*, Basic Books, New York, 1997.

Currier, D., Nelson R., *Dynamics of Human Biologic Tissues*, F.A. Davis Company, Philadelphia, 1992, pp. 9, 10, 258.

Deyo, R., "Low-Back Pain," *Scientific American*, August 1998, p. 98.

Dunn, Marcus B. H., Kampert J. B., Garcia M. E., Kohl H. W., Blair S. N., "Comparison of Lifestyle and Structured Interventions to Increase Physical Activity and Cardiorespiratory Fitness," *The Journal of American Medical Association*, January 27, 1999.

Feynman, R. P., *The Meaning of It All*, Perseus Books, Massachusetts, 1998, p. 15.

Frequently Asked Questions, www.post-polio.org, Gazette International Networking Institute (GINI), coordinator of International Polio Network (IPN).

Gavin, T., C.O., "Orthotic Treatment for Idiopathic Scoliosis: Current Concepts," *Backtalk*, published by the Scoliosis Association.

Gawain, S., *Creative Visualization*, Bantam Books, Inc., Whatever Publishing Inc. Mill Valley, Cal., 1978.

Gleick, E., "Land of the Fat," *Time International*, October 25, 1999, p. 2.

Gould, J. A., Davies G. J.(eds.), *Orthopaedic and Sports*, C.V. Mosby Company, Missouri, 1985, p. 88.

Gray, H., *Gray's Anatomy* 1901 Edition, Running Press, Pennsylvania, 1974.

Grimsby, O., *Clinical & Scientific Rationale for Modern Manual Therapy, MT-1*, Independent Study Manual, San Diego, 1998, *Neurophysiology*, p. 5.

Guyton, A. C., *Textbook of Medical Physiology*, W.B. Saunders Company, Pennsylvania, 1986, pp. 6-8.

Hayward, J. W., Letters to Vanessa, Shambhala, Boston, 1997, p. 7.

Hinkle, C. Z., *Fundamentals of Anatomy of Movement*, Mosby, Missouri, 1997, pp. 7, 82.

Holten, O., Faugli H. P., *Medisinsk Treningsterapi*, Illniversitetsforlaaet. 0608 Oslo, Norway.

Hoppenfeld, S., *Orthopaedic Neurology, a Diagnostic Guide to Neurologic Levels*, Lippincott-Raven Publishers, Pennsylvania, 1997, p. 73.

Humphrey, P. (ed.), *America in the 20th Century*, Marshall Cavendish, New York, 1995, p. 35.

Iyengar, B. K. S., *Light on the Yoga Sutras of Patanjali*, Thorsons, California, 1993, p. 1.

Jacobson, F., "Medical Exercise Therapy," *Norwegian Journal of Physiotherapy*, 1992: 59(7):19-20.

Jensen, M. C., Bryant-Zawadzki M. N., Obuchowski N., Modic M. T., Malkasian D., Ross J. S., "Magnetic Resonance Imaging of the Lumbar Spine in People Without Back Pain," *The New England Journal of Medicine*, July 14, 1994.

Jonck, L. M., van Niekerk J. M., "A Roentgenological Study of Motion of the Lumbar Spine of the Bantu," *Journal of Laboratory & Clinical Medicine*, June 1961, p. 68.

Justice, B., *Who Gets Sick*, Peak Press, Putnam & Sons, New York, 1987, p. 123.

Kabat-Zinn, J., *Wherever You Go There You Are*, Hyperion, New York, 1994, p. 45.

Kandel, E., Schwartz J. H., Jessell T. M., *Principles of Neural Science*, Appleton & Lange, Connecticut, 1991, pp. 14, 183-184, 390, 534, 761.

Kottke, F. J., Halpern D., Easton J. K. M., Ozel A. T., Burrill C. A., The training coordination. Archives of *Physical Medicine and Rehabilitation*, 1978, 59: 567-572. M. Trew and T. Everett (eds.), *Human Movement*, Churchill Livingstone Inc., New York, 1997, p. 93, cited by second source given.

Lamb, D., *Physiology of Exercise*, MacMillan, p. 260.

Langer, E., *Mindfulness*, Addison-Wesley, Reading, Massachusetts, 1989.

'Low-Back Pain,' Richard A. Deyo, *Scientific American*, August 1998, p. 3.

Malone, T., McPoil T., Nitz A., *Orthopedic and Sports Physical Therapy*, Mosby, Missouri, 1997, p. 539.

Myss, C., *Anatomy of the Spirit*, Three Rivers Press, Crown Publishers, Inc., New York, 1996.

Nitz et al., Brownstein, B., Bronner, S. (eds.), *Functional Movement in Orthopaedic & Sports Physical Therapy: Evaluation, Treatment & Outcomes*, ed., Churchill Livingstone Inc., New York, 1997, p. 147.

Olsen, A., *BodyStories*, Station Hill Press, New York, 1991, p. 51.

Ornstein, R. and Thompson R. F., *The Amazing Brain*, Houghton Mifflin Company, Boston, 1984, p. 27.

Parks, B., The myofibroblast anchoring strand—the fibronectin connection in wound healing and the possible loci of collagen fibril assembly. *J Trauma*. 1983;23(10):853-862.

Perez-Christianens, N. Etre d' Aplomb, Institut Superior d' Aplomb, Paris, 1983.

Rama, S., Bullentine R., Hymes A., *Science of Breath: A Practical Guide*, Himalayan International Institute, 1979, Pennsylvania, p. 36.

Schommer, N., *Stopping Scoliosis*, Avery Publishing Group, New York, 1991, p. 6.

"Scoliosis: Diagnosing Girls at Risk," American Academy of Orthopaedic Surgeons, September 1, 1994, p 1.

Silva, J., *The Silva Mind Control Method for Getting Help From Your Other Side*, Simon & Schuster Inc., New York, 1989.

Silva, Mira, Mehta S., *Yoga the Iyengar Way*, Alfred A. Knopf, New York, 1990, p. 8.

Taber's Cyclopedic Medical Dictionary, p. 159.

Taylor, C. R., Heglund N. C., Freeloading Women, *Nature*, 4 May 1995, p. 375.

The American Heritage Dictionary of Science, Barnhart Books, 1986.

"The Senses," *U.S. News and World Report*, August 1999, p. 5.

The Visual Dictionary of the Skeleton, First American Edition, DK Publishing, New York, 1996, pp. 42-43.

Thusius, A., Couch J. "Living on the Axis," *Yoga Journal*, July/August 1991, p. 71.

Twomey, L. T., Taylor J. R., *Physical Therapy of the Low Back*, Churchill Livingstone, New York, 1994, pp. 58, 66.

Vaughn, F., M.D., interviewed by Lisa Ann McCall, April 2000, Dallas, Texas, phone conversation.

Wallace, B., interviewed by Lisa Ann McCall, June 2000, Dallas, Texas, phone conversation.

Weil, A., Breathing Tapes.

Williamson, B., M. S., interviewed by Lisa Ann McCall, August 1999, Dallas, Texas, personal.

Wolff, J., *Gestezkter Transformation der Knocken*, Berlin, A. Hirschwald, 1884.

Wolkomir, R., "Charting the Terrain of Touch," *Smithsonian*, June, 2000, p. 44.

Woo, S., Buckwalter, *Injury and Repair of the Muscoloskeletal Soft Tissue*, 1987 Workshop, American Academy of Orthopaedic Surgeons, pp. 428-429.

Wyke, B., 1979; Freeman and Wyke, 1967, *Manual Medicine*, Thieme Medical Publishers, New York, 1990, p. 36.

Wyke, B., Presentation at APTA Conference, 1974.

Picture Credits

THE HARVEY CAPLIN PHOTOGRAPHY COLLECTION 1940s Zuni Olla Bearers, Zuni Pueblo, New Mexico, pp. 10, 47 / JUDY WALGREN boy with camel, Rendille, Northern Kenya, p. 13; Dinka boys with sacks on head, Southern Sudan, p. 16; women braiding hair, Rendille, Northern Kenya, p. 20; Indian women with baskets on head, Pushkar, Rajastahan, India, p. 26; Silhouette of man standing, Southern Sudan, p. 40 / BRUNO BARBEY man from Morocco, pp. 23, 106 / LINCOLN WILLIAMS man carrying load on back, Nepal, p 32 / THE IMAGE BANK child from Bali, p. 36; woman from Bali, pp. 36, 44; Men towing boat, China, p. 64; girl carrying sticks on head, Guatemala, p. 65; Beijing Opera, China, p. 195 / MARC ROBBINS Flamenco dancer Conté de Loyo, p. 158 / PHIL HOLLENBECK Flamenco dancer Conté de Loyo p. 38, 170; Janet Mwalalino from Malawi, Africa, pp. 31, 88, 194; African dancer Djely Moussa Diabate from Guinea, West Africa, p. 48; Belly dancers Nina and Cathy, p. 69, Belly dancer Nina, pp. 100, 200, 201 / LISA ANN MCCALL Portuguese fisherman pp. 64, 104; Portuguese fishermen, p. 210 / WENDY JAMES Devon James with reeds pp. 65, p. 110; Devon James seated on limb, pp. 165, 179 / FREDERICK FURSMAN 1874-1943 "In the Garden," p. 89 / JOAQUIN SOROLLA "Louis Comfort Tiffany," p. 94 / MARY VAN ARMSTEAD woman carrying baskets, Bangkok, Thailand, p. 113.

MODEL CREDITS Jody Baird, Robert Kane, Scott Herrera, Rachel Kane, Clinton Trammell, Jennifer Polavieja, Ginger Vitus the cat, Devon James, Phil Hollenbeck, Jennifer Parigi, Larry White and his dog Jake, Scott Drake, Melissa Myers, Djely Moussa Diabate and Conté de Loyo.

"ACKNOWLEDGMENT" ILLUSTRATION Gabriela, Yasmine, Moses & Victor.

INDEX